The New Face of Your Brand

Social Brand Advocacy grows Business with Customer Community and Co-Created Marketing

by Shawn M. Miller

ISBN: **978-0-692-77281-2**

Table of Contents

Foreword:

When social media burst onto the scene in 2006 or thereabouts, it was considered many things. It was a pleasant distraction. It was a new and novel way to make friends and keep track of family. It was an interesting way to share pictures and stories with more people. And, it was a time sucking black hole.

It certainly was not for business. Arguments ranged far and wide with all kinds of pundits espousing multiple reasons why social media would not be an effective addition to the business arsenal.

There were a few voices and early adopters who realized even at the time that eventually the ability to reach people in this new fashion would be valuable.

In retrospect, we laugh at those who couldn't see how powerful social media would eventually become. It has been years since we heard the slightest whisper from those arguing it has no place in business.

It is now a given that an effective social media strategy is an essential component of every business if you want to stay viable and grow.

Like everything else in the digital age, this happened at digital speeds. The number and purposes of social media have exploded, morphed and vanished as they wandered all over the virtual landscape.

Various platforms have changed from banning advertising outright to embracing advertising as their sole means of income. Changes continue to happen at breakneck speed.

It is now 2016, ten years after its "beginning," and social media is changing again and changing dramatically.

Just like earlier, most can't see it and consequently sit on the sidelines and implement what they can while they lament about the "good old days" of social media. Whatever that means.

Like those lone voices of yesteryear who understood early on the power that lay in the structure, communication style, revolutionary technology and organic development of social media platforms, there are a few today who understand the landmark turn that took place in 2016.

Shawn Miller is one of the prophetic few at the absolute forefront of the monumental sea change that has taken place in social media.

One of the amazing pronouncements is the recent death of "Organic Reach." With this funeral, so begins a saga of the modern version of "word-of-mouth advertising."

Unlike the days of yesteryear, word-of-mouth advertising is not slow and occasional. It takes place at light speed and multiplies like a virus. The new trend that the wise ones see and are adopting is the creation of social advocacy communities.

Shawn brilliantly makes the case for this massive shift and, far more importantly, explains exactly how to take advantage of these changes in your business.

Just a few short years from now, we will all be looking back and wondering why we could not see the blindingly obvious at the time.

Those who either see it now, or have the good sense to take advantage of the vision of those who do, will be sitting on fat bank accounts, dominating market shares and a position of leadership in their market segment.

Well written and easy to understand, this book is a landmark in the evolution of social media. I would be remiss if I simply said "You need to read this book!" Instead, I will say it this way: "You would be extremely unwise to miss this opportunity."

Kellan Fluckiger
CEO: Authority Institute
Founder: Break the Cage NOW

About the Author:

Shawn M. Miller

Chief Experience Officer, Smync

www.Smync.com and @gosmync on Social

Shawn M. Miller is an **Entrepreneur, Investor and Advisor** with over 20 years of international business experience specializing in marketing, customer service and training. As Chief Experience Officer at Smync; Shawn is leading development of User Experience on both the brand and the customer side of Smync hosted Social Brand Advocate Communities.

Shawn lives in **Holland, Michigan, USA** with his wife Amy and two teenage daughters. The Millers are very active in their community and enjoy the outdoors in Michigan year-round.

Klout tracks Shawn among the **Top 0.1% of Influencers on-line** on the competitive topics of **Business, Entrepreneurship, and Social Media**.

Connect with Shawn on https://www.linkedin.com/in/shawnmmiller

on Twitter @ShawnMMiller

or Snapchat username: theshawnmmiller

Introduction

This book is written specifically for those responsible for the marketing of consumer brands, though leaders of any business will certainly benefit and should pay careful attention.

My interest here is to give insight, tools and motivation that will help you grow your company rapidly, leveraging a combination of emerging consumer trends.

In a recent conversation with Jeffrey Hayzlett, a friend and advisor to our company, Jeffrey made it very clear that in today's marketplace, any attempt to separate the marketing of your business from your consumer's experience with your business is absolutely doomed to fail.

Amazon CEO, Jeff Bezos, recently said, "you don't control your brand anymore. Your customers do. Your brand is what other people say about you when you're not in the room."

Social Media thought-leader Ted Rubin recently expounded on companies' ineffective focus on Recognition vs. the more impactful Reputation. The message is clear: our culture has shifted and your marketing must shift with it.

This book is the first of its kind. It teaches brands to empower the next generation of customers so they become your digital marketing department. It will help you leverage their addiction to mobile and social media so they sell your brand for you.

Culturally, we are in the midst of the massive momentum swing from a 'Me' centric culture toward a 'We' centric culture. Research shows this shift is about one-third complete. This fact, combined with the ever-growing consumer influence wielded by two generations of customers that are digital natives, requires that brand marketers look at their craft in an entirely new way.

These digital native consumers have such disdain for your marketing that they oppose its very presence on their devices. On the other hand, this disdain drives a strong sense of obligation to help each other make good buying decisions with companies that consumers feel are trustworthy.

These shifts have given rise to a new branding 'Axis of Evil.' It is the intersection of Distraction, Disconnection, and Distrust. This Axis threatens the very way of life for your business.

To effectively combat this Axis, you are called to trigger the end of any 'Us vs. Them' mindset in the culture of your company, then replace it with 'Community and Co-creation' of your brand which includes your customers' participation.

In 2016, less than 50% of those companies listed on the Fortune 500 in the year 2000 remain on the list today. It is critical for the life of your business that you consider the subject matter discussed here and take action immediately.

On the other side of this change is an amazing opportunity to leverage a marketing modality which is as old as, "this apple is pretty good, you should try it." Word-of-mouth. Social Media networks have jacked up word-of-mouth recommendations as though they were on steroids.

This has come to a place that none of us, marketers or not, could have ever imagined. Social Media is truly a revolution in human communication. If we are astute and act quickly going forward, we get to be participants in it and shape how it evolves with respect to business.

The opportunity exists today to empower your own customers, employees, vendors, friends and stakeholders to do the very human activity of sharing, which we know as word-of-mouth, in a new, powerful and far-reaching way.

These principles will leverage the omnipresence of social media and mobile devices in our lives as a new form of marketing. We will call this new power 'social word-of-mouth' or 'Social Brand Advocacy,' interchangeably.

If you have any percentage of your existing clients who:
1) would recommend your products and services to a friend or family member
2) would be willing to do so publicly

3) are active on social media (as most people are)

you're sitting on a gold mine and probably don't even know it. So let's see how big that gold mine is for you.

The **New Net Promoter Score for 2016** and beyond is defined by the relationships between these three numbers.

Start by estimating the answer to these questions. How many of your customers:

1) Would recommend you to a Friend?
2) Would do so publicly?
3) Are active on Social Media?

What percentage of your existing customer base does that represent? How many customers would meet all three? What if you added to that number many of your employees, close colleagues, vendors, and even friends and families related to your business? What would that number be?

What if I told you that on Facebook today, each act of sharing and recommending posted by each of these friends and family, your customers, and more, could influence 40,000 others?

Think about that!

With only 25 engaged and empowered Brand Advocates, leveraging the principles of social brand advocacy, you can have the equivalent effect of consistently placing your brand message in front of 1 million people! All this with zero advertising budget required.

Using a simple correlation to Earned Media Value of Pay per Action, those 25 would trigger the equivalent of $3.3 million in ad spend value. But it gets even better...

Remember, we're talking about the direct connection of actual people talking with actual people with whom they have real relationships. Therefore, the sharing of your brand story is delivered by someone they Know, Like, and Trust.
Because of this advantage, your brand story will immediately pierce the Axis of Evil and provide approximately a 70X return on investment when compared to that of paid advertising.

Messages delivered by enthusiastic 'Brand Advocates' come into the consciousness of the recipients:
1) in context
2) at an appropriate time
3) with almost complete trust
4) with little need to fight the "battle of overwhelm" that prevents our advertising messages from reaching our audience.
AND, they can do this all day, every day.

This is Social Brand Advocacy, word-of-mouth marketing on steroids because of social media!

A mere 12% increase in this activity on behalf of your brand will double your present revenue growth rate! This is an activity that many, many people are likely already doing on your behalf, are excited to do, and would like to do more of, if you would simply empower them to do so.

This will cost you just a few nickels for the technology and tools required to make this run well.

Most importantly, it will require a significant commitment to this reality: **you cannot segregate any aspect of your customer's experience with your brand from your brand marketing.** You cannot separate people's social media sharing and expression from their customer experience. This trend will only accelerate and get more powerful going forward.

This is a significant shift and a major commitment for businesses. But it is one that, if we're being realistic, we see clearly must be made. Any company which fails to make this shift as we build momentum in the 'We' culture and account for the economic dominance of Millennial and Gen Z consumers will fail!

As an example, we can see the glowing brilliance and global domination of Mr. Bezos' company, Amazon, which grew from its start to today on exactly the Brand Advocate, word-of-mouth, principles that we're discussing here and throughout this book.

While I will list for you research references that I have relied on, this will of necessity be limited due to the exponential nature of technological change and a rapidly changing marketplace. Most of the available research that academic readers might wish for is at minimum two years old.

Analysis of that data is also at least a year old. By today's standards that's the Stone Age. The lack of hard data may offend a select few. This book is not written for them. This book is written for those who understand that the marketplace and social media are changing too fast to wait for that analysis.

A healthy dose of intuition is required to identify and make the changes that will let you keep pace. Besides, keeping pace is not your objective. Leading and market domination is where those at the edge want to be.

Combining the reality of today's fluid marketplace and the exponential participation and omnipresent expansion of social networks requires that we interpret and act rapidly on what is present today. These facts and trends are significantly different than they were just one quarter ago. The pace of social expansion and change will not slow, but rather accelerate. All marketers need to accept these facts or be left behind. We can't afford to get stuck on specific details of the minutia of the day. Rather we must simply accept those things that are effective today and use them today.

We seek also to leverage basic principles of human behavior which drive social media and fuel social brand advocacy. At a recent large Social Media Marketing conference in Chicago, I found that the vast majority of those responsible for the day-to-day actions of marketing for the best and brightest companies are mired in minute details of tactics.

They have lost sight of the fact that on the other end of Social Media are actual human beings like themselves who are there simply to communicate with one another to share and enjoy and be entertained as part of social networking. This is greatest revolution in human communication since the invention of language.

You must release your mind for a little bit here and consider some fundamental basic truths about how people interact. We have to understand why they do what they do, and how they do it effectively.

We must discover how we can use the tactics and tools of social media to empower and motivate them to interact on our behalf. Better yet, advocate for our collective behalf, because we've become members of the same team.

Our job is to grow our company and be profitable, using the best available of whatever the marketplace happens to be today. That is exactly what I provide you here. These are the strategies and the tactics which will transition your business profitably into the digital native (Millennials and Gen Z) generations' culture of consumerism.

If you discuss the concepts in this book with a 60-year-old, you may get confusion and condemnation. Discuss them with a 20-year-old and instead you'll get, "Yes! Exactly! Why don't they get it?!"

By the way, I am a Gen Xer and you may be too. If that's the case, I have great news for you. Your unique ability to understand both the confusion of the 60-year-old and the enthusiasm of the 20-year-old are a blessing for you personally.

Your new role in business may be to bridge that gap of understanding and be the project manager of these changes.

Note from the Author:

The book is the direct result of my pouring myself into this space, having materially engaged with Smync and seeking the best solutions that I could compile for brands to bridge the trust gap and make Social work as intended.

My whiteboard and sketchpads are my 'Zone of Genius'. The places where my right brain dominant mind figures things out and the places where I best communicate those thoughts with others.

So, while I know this is unorthodox, the Illustrations here are all hand drawn my me in the same way that I would if we were in the room together.

I hope you appreciate my style and enjoy this personal connection:

Chapter 1
Here lies 'Organic Reach' R.I.P.

The year 2015 is widely recognized as the year that organic reach died. *Organic reach* is the total number of unique people who were shown your Post through unpaid distribution. Paid **reach** is the total number of unique people who were shown your post as a result of paid advertising.

In June of 2016, Facebook released a News Feed Values Statement regarding prioritization of content in individuals' News Feeds. Facebook then delivered statements detailing how content is prioritized in your News Feed.

The signs are very clear for publishers and brands using Facebook for marketing - this update is shoveling more dirt over the coffin of organic reach. What Facebook clearly laid out is that content from "Friends and Family" is #1, News is #2, Entertainment and brand content, including Facebook Live, is #3.

While it has been dead for a while, this clearly states Organic Reach for Brands on Facebook is DEAD, BURIED, and GROWN OVER! It's time to stop mourning and move on to something that will work going forward. Forget about hoping for a resurrection of Fan Page generated organic connections driving business from Facebook.

This makes it crystal clear that Brands have only two viable options to reach consumers on Facebook: 1) Social Brand Advocacy, which is highly effective and offers an unmatched ROI, or 2) Paid Advertising.

The best answer – and the top of Facebook's priority list – are the real people who have the trust and attention of their friends – and priority in the content in their news feed.

These trusted people are the catalysts who have the ability to Stop the Swiping (the ubiquitous sideswipe motion that means you have already lost attention) and get focused attention from your desired audience.

This is the hidden gold mine - the customers, fans and stakeholders of your brand communicating naturally on Facebook.

Facebook was very clear in their goal statement about 'Authentic Communication.' This change now requires Brand Advocate intervention/representation on behalf of your company.

Paid Access
Organic reach didn't die by itself. They killed it, intentionally. Facebook, followed immediately by others (including Instagram, which Facebook owns) made a series of strategic changes of the algorithm that dictates the appearance of your social posts. These changes began to intentionally take away your ability to reach the audience that you worked so hard to build.

What marketers unknowingly did in building vast followings on social networks, was insert a gate that they did not control between themselves and their rented customers and prospects. They were never 'your' people. They belonged to the network. The rules controlling access to the data about them and your marketing to them were unilaterally changed.

This was done for the same reason that you do your marketing to customers and prospects. The social networks saw an opportunity to make money. The opportunity for Facebook to create maximum profit from their platform and make you pay to access your own clients and prospects became glaringly obvious. They began to make that shift in 2014.

Most of you recognize this access issue now, but I've discovered that you don't recognize that you have endangered your community in two ways. First, when you rented access to those customers and prospects, it did not occur to you that the rules might change dramatically.

Like everyone, you relied on the network's ability to communicate with your people. The networks' control over the environment and your ability to create your own community was not a problem. Now that the rules have changed, you have likely not yet tried to move them into a community that you control. For now...

Open Source Lists

The second problem is one that is almost totally unrecognized by business owners and marketers. Frankly it is one that scares me more than the first. I understand the networks' motivation and expect to pay to use their platform. It is their platform and their right, and their purpose is to profit from access to those people on that platform.

The more concerning element is that, by maintaining the community for your business on that social network, you are exposing your very best clients and prospects to the possibility of being pirated by your competitors. Your fan base on a network is an open source list.

This means that any competitor wishing to have exposure to your best clients and prospects and to replicate your ideal client avatar is free to access that information to an alarmingly specific degree via your social community as it is currently maintained on the network.

We'll talk more later about the capacity to own your own branded community and the critical nature of this move for the long-term sustainability of your business.

Information Overwhelm
The second factor in the death of organic reach is the sheer overwhelm of content that is burying every consumer. The advantage of the Information Age, combined with the massive focus of 'creating content' has built a monster with a present focus on non-sales-like, content marketing.

Just two weeks ago I was at Google in Chicago with the CMO's of some of the largest consumer brands in the world. One of the startling statistics shared by Google was that 400 hours of video content is being loaded to YouTube for each minute of time going on the clock.
Every minute. Just YouTube. Not counting everything else. Talk about overwhelm!

400 Hours of Video for every 1 Minute of Time!

Filtering that unimaginable quantity of content for reasonable functioning of search engines and social newsfeed requires that they create ever more effective and selective algorithms.

To satisfy consumers, these algorithms must ensure that only the very best, most engaging content is served as correctly and as personally as possible to each specific user. This is a daunting task.

I'm glad I'm not responsible for it. It is a task, however, that you must comprehend if you want to understand why organic reach is dead and seek instead viable alternatives to reach your customers.

Chapter 2
Axis of Evil – Distraction

Imagine standing on one of the striped lines on the 405 freeway in Los Angeles at 5 o'clock in the afternoon as thousands upon thousands of cars buzzed past you as quickly as they possibly can. Form a vision in your mind of your promotional social post as one of those vehicles flying by your prospect in the midst of total overwhelm.

Now you begin to understand the volume of noise and overwhelm that content marketing strategies have fueled. We've oversupplied our own inability to connect with our intended audience.

Many top marketing gurus like to suggest that today, you have perhaps 6 seconds to capture the attention of your prospect. I maintain that is a vast overestimation of the time you have, at least on Social Media.

Swipe, Swipe, Swipe...
(I couldn't help imagining replacing these three words with 'wipe, wipe, wipe,' given the following imagery.)

To be real about the scenario on social media with content, even paid content, imagine yourself (if rather ungraciously) sitting on the toilet, cell phone in hand, swiping through a Newsfeed for some eye-catching name, video movement or image that causes you to stop just long enough to have a look.

A short look!

Jay Baer says, "You're not just competing with other brands for their attention, you're also competing with friends, family, music playlists, soccer games, breaking news and nights out on the town." It's a noisy Newsfeed and the Network has the deck stacked against you unless you pay and pay dearly.

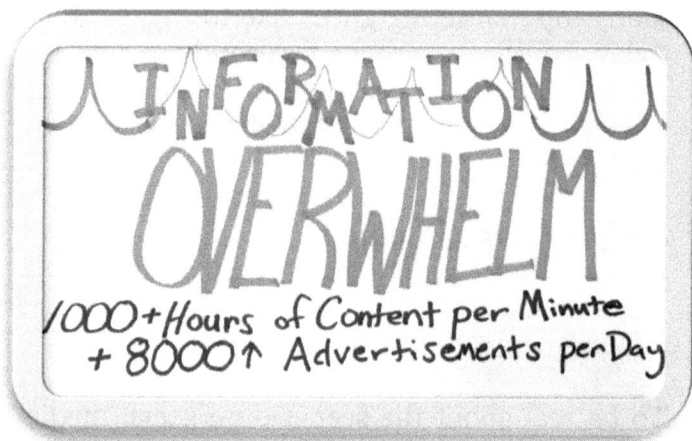

INFORMATION OVERWHELM
1000+ Hours of Content per Minute + 8000↑ Advertisements per Day

How to Stop the Swiping

The better answer is right under your nose. You're ignoring the people who have the ability to gain that **Stop Swipe** attention from your audience. This is the Goldmine you are sitting on that I talked about earlier.

It is customers and friends of your brand and their close relationship to new prospects. Brand Advocates provide trusted, relevant message delivery that drives sales. They amplify your brand story for little to no cost. They can and will put your message into the hands of your consumers. Remember what you learned earlier about who controls your brand message!

Advocacy empowers your customers to deliver your message right into the eyes of an ideal prospect at the ideal time. Cutting through the noise and overwhelm because it's coming from someone who has the ability to get mind capture. Fred Reichheld—the father of Net Promoter Score—says: **"The only path to**

profitable growth may lie in a company's ability to get its loyal customers to become, in effect, its marketing department." This is why I maintain firmly that social brand advocacy is the holy grail of marketing today, not more content and not any magical noisy 'Influencer.'

Growing your company's engaged Brand Advocates by just 6% can 4X the Revenue Growth rate of your consumer brand. Effective brand advocacy activity from just 6% of your client base will increase your market share by nearly 10% and at a much more profitable operational cost than your competition.

Accept the fact that organic reach is dead and stop ignoring the people who have the attention of your ideal clients and prospects.

Chapter 3
Axis of Evil - Disconnection

When most non-millennials think of Millennial's, they have the wrong picture in their mind. Millennial's are probably older than you think they are. The oldest millennial has now reached age 35 and they make up the largest consumer spending power group in the United States.

They're just not just kids anymore! Along with the Generation Z which comes behind them, they are a new culture, one that we collectively referred to as Digital Natives. They have been raised with full time Internet and device access to entertainment and communication.

Technology is an inseparable part of their life experience. And so is advertising. They have seen advertising since they were old enough to focus on a screen. And they're thusly immune to advertising, when they are not annoyed by it.

There has been an underground phrase in Social Media for years that 'when the Marketers move in, the people move out'... A recent Harris Poll commissioned by Lithium Technologies found that 74% of digital natives, (comprised of Gen Z and Millennials,) object to brand advertising being in their Social Media Feeds at all.

An astounding 56% of that population are cutting back or jumping platforms due to the presence of advertising. World-renowned PR expert Richard Edelman stated: "90% of marketing budgets are currently pointed at ineffective and unsustainable advertising and promotional campaigns." This is a mission-critical problem that you and I must solve to stay in business.

On the one hand is the marketing perspective that social media represents a mass gathering of likely prospects and customers. On the other hand, those very customers and prospects demonstrate a clear feeling that social media should be an ad free zone.

Millennial consumers also demand that companies deliver something extraordinarily different.

"Loyalty has changed overnight," said Ken Clayton, VP of Dassault Systems. "Customers have gone in a direction for reasons we didn't understand. And now the experience has to speak through our products. Customers expect a personal experience," (with integrated shareability.)

We're on Social Media to connect with our friends and family. We're on Social Media to be entertained. We're on Social Media to hear the news and see what's going on. We're not on Social Media to be marketed to.

And now, the ones with the buying power are no longer willing to submit to the 'deal' that in order to receive high-quality content they must be subjected to unwanted advertising.

Ad Blockers
Adding to the trouble of consumers willing to leave a platform due to advertising, another glorious opportunity for prospects is the use of software to block your ads. We called them Ad Blockers.

The use of ad blocking is radically on the rise. Marketers are still figuring out what to do about the trend. Consumers using ad blockers over the course of 2015 grew by nearly 200 million.

This action cost marketers an estimated $22 billion in lost revenue. By 2016 it is estimated that ad blocking will cost publishers nearly $41 billion.

The trend will continue to grow. Currently 22.5% of web users are using ad blocking software, and the reporting shows a growth rate of 43% per year. This is driven primarily by an annoyance with advertising.
It is also significantly fueled by concerns of safety and privacy. People increasingly seek to avoid being tracked and monitored against their will when they spend a quarter or more of their lives online.

You can break through, of course, if you spend more. You can always spend more and more on paid advertising, but paid reach is very expensive and at best, limited in effectiveness. Only 1% of digital native millennial's self-reported being influenced by advertising.

It is obvious that Brand Advocacy is the way to avoid all of this waste and turn Likes into leads and Social activity into Sales. We trust our friends, family, colleagues and coworkers as this is the only way to reach history's most jaded consumers.

Fueled by the global reach and 24/7 nature of social media, word-of-mouth is no longer limited to immediate circles, geography or time zone. The Goldmine is just getting shinier as we explore it!

For the cost of one 30-second television ad, you could fund an aggressive Advocacy program for many years. Advocates are more effective, more credible and far more sustainable than any paid advertising we could possibly dream up.

Advocates know the relevant context of the struggles and the specific timing of consumption-driven life changes of your ideal consumers in ways you can only imagine.

Chapter 4
Axis of Evil – Distrust

According to the latest research, American consumers believe that marketers act with integrity, a mere 4% of the time. Ouch, there is a gut punch, they trust us less than Congress!

And as just revealed, they trust advertising only 1% of the time. Here's the paradox: You're dying to have your content deemed sharable by your customers, but they don't perceive your brand as a trusted source, so they won't share with people they feel are depending on them.

The Likelihood of a Share ↑ The Reputation of the Source

It would risk their own social standing and personal brand growth! Bryan Kramer's Shareability Quotient says that the likelihood of a share must be greater than or equal to the perceived reputation of the content's source.

For them to want to share your stuff, they need to trust you and identify with the content you're producing, period.

However, more than 9 in 10 indicate that the recommendations of friends are the most trusted source of information about purchase decisions.

Therein lies the trust you are seeking. And nearly 90% of all B2B purchases are begun with a personal referral. This isn't a B2C or B2B thing, it is a human thing. Conveniently for us who are excited about Brand Advocacy, 74% of Millennials report feeling **that it is their duty to help their friends and family** make better buying decisions.

Advocacy, not Advertising drives purchase decisions today.

"I share to enrich the lives of those around me."
– Consumer Insight Group study participants
In order to benefit from that massive trend, as brands we must create scenarios in which we warmly invite people into engaged relationships with us. The empowered buyers of today

demand a new level of customer access and connection.

"94% carefully consider how the information they Share will be useful to the recipient" - Consumer Insight Group study

Humans are designed for tribes and codependence, that's where we thrive and function our best. It is where we let our guard down and connect.

By practicing the actions of bringing together your company and your customers you will create an interdependent culture of connection which will fuel this highly effective and infinitely sustainable funnel of community in context around the problem that you solve.

Your actual results in defeating the Axis of Evil with Brand Advocacy will be difficult to precisely predict without knowing your specific business. Trends indicate that the higher the price and more exclusive the product, the greater the impact Brand Advocacy will make.

Regardless of your product or price point, I can assure you without question that effective Brand Advocacy will provide a multiplication of both your growth rate and your profit margins. Trends also indicate that it can be triggered and active in the market very quickly.

This is not surprising given the global and instantaneous reach of Social Media. This Goldmine is real and is present. It is yours to reveal, not by digging, but by removing barriers and inviting connections.

Chapter 5
The end of Us vs. Them Marketing
(aka: Combat with your Customers)

Words have power. Creative power. God spoke the world into being. With our language, we fuel our thoughts which fuel our emotions, which fuel our creation. This applies to your company and your company culture. Your company is simply a gathering of people, and as you now understand, it also intimately includes your customer base.

Your business is not one-sided anymore. The defeat of your Axis of Evil means the inclusion of your customers to achieve rapid and sustainable business growth. Your understanding of the power of the language you use gives you power to create the culture of your company.

As a leadership group at Smync, we strongly Advocate for the end of military and combat language in all marketing and sales to our customers.

This will bring the end of Us vs. Them Marketing. The use of military/combat language is pervasive in marketing. It seems to be especially prevalent among those who profess to be believers in Advocate marketing.

One leading expert in the field recently referenced in his national level keynote an 'Army of Zealots.' While we can understand the sexiness of that headline, and a desire to be controversial, you don't want to wage war on your customers via your Advocate Community. You are not seeking to outsource Advocacy to mercenaries on your brand advocacy platform.

We want sustainable relationships. As people and consumers ourselves, we do not desire to have an army engaged in a campaign to breakthrough, overcome and hammer us with a business deal either.

We must end the Us vs. Them mentality for the future of our business, and it begins with our language. This is simply not what we want for ourselves, our families or our own company and so we seek to end that culture with the language we use. Instead, we seek to create a culture of cooperation, inclusion and excited participation among our most valuable asset, our clients.

Co-Creation with our passionate, loyal, customers!

Consider carefully your language and take it to another level. Stop thinking about customer conquests, churn, up-selling and closing. Instead think conversation, referrals, ownership, and cooperation.

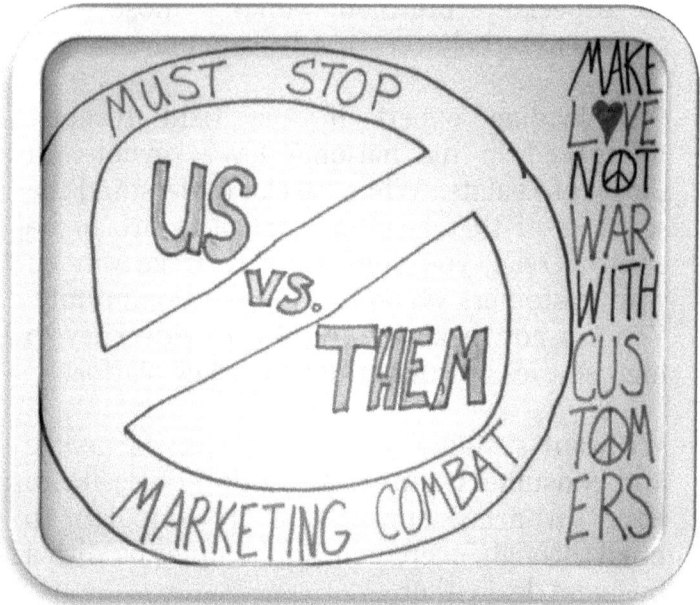

Combat-won customers are not sustainable customers. People trust Advocates because they know the advocates have nothing to gain. Advocacy is actually the highest level of relationship.

Most would say that loyalty was the highest level of relationship, but that is not so. Many consumers remain loyal to a product because it is convenient or price effective or the only option.

That is quite different than proactively recommending it to people they care about. It's a whole new level for a customer to actively go out of their way to engage their friends and family and recommend a product or service. Advocacy is not a short-term combat strike for quick sales, this is a transformational customer engagement and empowerment endeavor.

These people will have your back for years! They are an asset of your business.

There are about 500 billion social word-of-mouth impressions each year, just in the United States.

Do not fight the empowered customer, befriend them, harness them, integrate consumer community and your company as one group with co-objectives.

Presently, about one quarter of American adults are Brand Advocates online. But, more than one half of consumers report that they are highly likely to recommend if encouraged and empowered to do so.

Holy cow!

Chapter 6
'We' Culture and the Modern Consumer

Social media is a true revolution in human communication. All previous advancements allowed for greater effectiveness in transferring one message to many people. From stone tablets, to scrolls, radio, television, signs; all were developments that magnified passing information from one source to many, as an audience. But not Social media. It is truly a revolution in communication and one which I have referred to continuously for about 10 years now as the People's Media.

With the power of this People's Media, there is now no limit to the viral nature and exponential growth potential of word-of-mouth marketing!

This technology driven communication is further fueled by our state of being only 1/3 of the way through a 40-year shift in culture from a 'Me' centric Culture to a 'We' Centric Culture. We are still on the momentum building side of that transition.

This cultural shift, along with the perpetual and growing presence of digital devices and social media, are the key cultural drivers of the Millennial consumer that your brand must learn to connect with and engage.

There are many complex factors to the 'We' culture and you can study them in detail in the book PENDULUM, if you wish. I recommend you do, but for now what you need to know is that, along with their desire to share and help one another in community, and their distrust of companies and advertising, the Millennial buyer in 'We' culture rejects all of the principles of brand positioning and marketing that we learned and perfected over the previous 40-year swing. Apologies to Robert Chaldini, but the gig is up, they aren't buying like that anymore!

Mainstream culture today will reject hype, arrogance, wishful thinking, falsely created scarcity, any form of sugarcoated bullshit and self-righteousness.

They even reject the idea of big dreams of any one person and prefer a 'show me' display of action. This culture will embrace: authenticity, teamwork, humility, action over vision, hard truth, and has an expectation of personal responsibility for service of the greater good.

This is now a 'talk is cheap, show me' culture, and this is just building momentum. No one is in a better position to show them than others they are naturally connected with both on and off line.

This trend is not limited only to Millennial's and Gen Z, but it is all that this particular group knows. It is how they will react and it is what will guide them in all areas of culture, including their consumer behavior.

These shifts require a change to an inclusive, 'Pull' style of marketing versus any coercive, 'Push' based marketing which may have triggered action in the past.

Push marketing will trigger action. It will trigger: ignoring you, turning you off, blocking you, protesting you and making an example of you!

In today's consumer mindset, people are striving for the very top of Maslow's hierarchy. However tenuous our current prosperity may be, (depending on your political stance) our recent economy means that our young people have never known any real hardship. They have no thought or concern for the low-end of Maslow's triangle and are clearly focused on the highest levels.

These levels drive us to make a meaningful impact on the world, to be part of a close and exclusive tribe and to have a high social standing in that tribe.

Today's consumers want to add to their social capital, their public reputation, just as my Gen X culture did, but they want to do it by helping as many people as possible make the greatest impact possible. The power of connection and recognition cannot be underestimated in understanding, motivating and mobilizing the digital native generations.

Note the amazing correlation between the motivation of the Millennial and Gen Z consumer, who longs to act as your Brand Advocate, (assuming they love you,) the We Culture as a whole, and the Principles defined by the Word-of-Mouth Marketing Association, WOMMA:
1) The resource must be credible, honest and authentic.
2) The resource must be respectful, transparent and trustworthy.
3) The resource must be Social, listening, participating, responding, and engaging.

4) The resource must be measurable, able to define, monitor and evaluate for accuracy and effectiveness.
5) The resource must be repeatable, over and over again.

Word-of-mouth is the ideal Marketing Channel for today's buyers and this momentum is only just beginning. Social Media is only getting more influential in their lives.

If you want people to trust you and share your brand story, you must:
1) work to humanize your interactions
2) empathize with your customers and demonstrate it
3) be transparent about what you stand for as a company
4) do the hard work of creating content that is sharable (or team up with some folks who can do this for you, including your own customers.)

Chapter 7
The end of Vanity Metrics

You can trade money for Likes, but you can't trade Likes for money and you can't put Likes on the Balance Sheet.

This statement is a painful reality for many marketers who have obsessively engaged in tactics and invested literally billions of dollars in creating social impression activity. They have sadly learned that this activity is effectively useless.

This wasteful behavior is now being taken to task by C-level executives and boardrooms all through corporate America who demand hard results for dollars used. It is also being thoroughly examined in the back office of every small business.

It's time to end the putting 'eyeballs on your logo,' Mad Men Style Marketing on Social Media.

You can waste a lot of money and grab a lot of Likes, but they are only vanity metrics. At best they are highly inflated reports of the real reach and influence of your attempts to connect with your ideal audience.

The perpetuation of this fallacy is driven by a combination of two things, neither of which are structured for your benefit. First, there is every social network's desire to increase participation and build revenue. Second, is the social media marketing worlds' desire to be validated in their activity and feel good about the impact they are making.

Neither one of these forces drive sales in any real sense. These vanity metrics are of no real value to anyone. Besides they give no meaningful representation of actual engagement or connection or relationship. Actual engagement with your ideal audience is all that matters. You'll never get to love if you trade in likes.

If you're a C-Suite leader reading this and you still imagine that these ephemeral statistics matter, you've got to let this go.

Marketing level employees I've met all over the country admit that they did such a good job in the past convincing you of the value of these metrics that you now won't let them off the hook for producing them.

If you won't listen to your own people, perhaps you'll listen to me. These 'impressions' reports are meaningless and requiring them is hindering your team from focusing on the engagement that will actually create relationships, make more sales and grow your company.

Most importantly, you need to establish a metric that matters to your bottom line. These can come in various forms. For example:
1) increased sales
2) increased lead generation
3) reduced customer complaints
4) building brand reputation

The goal must be tied to a numerical value, a deadline, and support your larger marketing strategy and corporate goals.

The results from these goals, when tied together across all your systems, can be productive. For example, one goal might be an increase in market share position by 3% in one year. Then your objectives clearly branch out from what is necessary to achieve this goal over that year.

Here are some other examples from proven programs:
1) Increased social engagement by X%
2) Increased social referrals by X%
3) Established a traffic conversion rate from social media to your own digital properties
4) Established email, or opt in, conversion rate

Clearly this will not be 100% accurate or successful in your first attempt, but it's critical and you must do it. Building your metrics from your objectives will allow you to measure how your social engagement links to your overall business objectives.

The Buzz on "Earned Media Value"

One of the latest marketing buzz phrases is 'Earned Media Value.' This can be a good metric that people should monitor and understand.

Earned Media Value – the dollar amount assigned to **media** that is gained through social endorsements – is a tangible way to measure the **value** of social word-of-mouth marketing programs. But it is important to take this a step further so that it does not become another vanity metric.

You must ensure you have the resources to connect tracking to how much business actually happens from your social media marketing efforts.

Ideally, you should understand how much you would have to pay for the actual impressions generated by your unpaid traffic (such as referrals from your Advocate community,) so that you can understand the sources and value of engagement and appropriate budget and allocate priorities properly.

Success in marketing cannot be measured by eyeballs, and we're starting to see the big shift on this, but it **will** take time. You're competing for trust and attention, not numbers of eyeballs. Relationships matter, engagement matters.

We're finally beginning to see even the top advertising gurus admit to this. Release your pride over vanity metrics and establish an approach to understand what you really should measure and what it really does mean for your business.

Earned Media Value should encompass:

1) Engagement - did someone DO something? Click, Visit, Sign up or Share?
2) Revenue - just because you "earned" a certain amount of visibility doesn't make it valuable...brands are in business for

business so, how did the action translate to business?

3) Loyalty/Trust - how high is the engagement and what is the sentiment of discussion around the brand? If a lot of people see it, but don't do anything, or worse yet, do or say something negative, what does that look like?

Earned Media Value is one of those abstract marketing concepts that employs its own special form of vanity metric voodoo. It drives sports sponsorships – naming rights to stadiums, logos on soccer and cycling jerseys – all under the premise that the sheer visibility of your logo (with a magical measurement of how many people either saw it on television, in the media via photos of the event or in person) equates to the value it would have cost to purchase those impressions and translates into cash.

Now, many talk EMV whenever content catches a bit of a social viral wave. Agencies and pundits are quick to talk about an astonishingly high amount of value received. There's one question to ask about all these impressions: **SO WHAT?**

Every day, we all see 80,000 or more brand impressions. We ignore 99.9% of them without consideration. Of the 0.1% we actually give attention to, depending on our trust in the source and the context, we do something with it. Maybe.

As a marketer, you are trying to build the connection between your company and consumers. There is no prize for making more noise – you're either paying for it, or justifying your measure of success by it, neither of which contributes to the success of your company. Social Impressions as a KPI needs to be killed.

Smync leadership has been proclaiming this for a long time and is excited to see a movement among thought leaders like Gary Vaynerchuk and others. Impressions don't matter. If you care about them, stop. If you're using them to justify your job, you're in trouble.

The number of impressions you think you're reaching is likely highly inflated anyway. (If there's anything less concrete than the value of impressions, it's the inferred impressions many market platforms use because they don't access the real data.) They can't get deposited into the bank account and they have no impact on your business.

I understand that without an impression, nothing else can happen. But what matters is what happens next – did somebody trust or find that impression worthy of engagement or action? Did they click, sign up for something, share it because they saw value in it for their friends? Did important content in context get viewed or better yet, did somebody actually buy something?

If the Chewbacca Masked Mom had posted a 7-second Vine of her laughing in her mask, many people might have seen it; laughed, liked it and that's it. But because she put everything in context – where she got the mask and the pure joy she got from it – the 'Earned Media Value' became clear and measurable. She drove traffic to Kohl's websites, she drove searches for the mask, and most importantly she drove sales until there were no more masks to sell. Real impact, real results, real business.

MEASURING EARNED MEDIA VALUE

As a marketing professional, you must realize "the noise" is a constant – billions of impressions per day. There are so many that whatever impressions you garner aren't a measure of success, they are a cost of doing business, an expense – not a value.

You are competing for Trust and Attention. While both of those elements are very real, their measurability is challenging. To succeed, you need to have both before consumers do something that impacts your business.

In measuring Earned Media Value today – put the following components into your calculation:

Engagement – someone taking an action shows value. A click to a digital property, a social share, a check-in at a physical location: all things that say seeing a piece of content drove an actual behavior that relates to your business.

While some might say that you can't put a social share on the balance sheet, there is an implied endorsement that comes with sharing. You're telling your own friends and family that you viewed this as worthy of sharing. This implied endorsement is why people will re-share it up to 24x more and it can trigger a purchase up to 50x more than a paid impression.

Lead Generation and Conversion Goals – Have a new product that came out and people are liking, sharing, and commenting about it on social media? Great, but how many of them actually clicked through and viewed the content that they are raving about?

If people go from social media to signing up for your email list, that's showing interest, that conversion provides a value.

Sales – You know, the good stuff. If I'm the owner of a local coffee shop and some guy starts juggling battle axes lit on fire while playing Blues Traveler on the harmonica on the street in front of my shop, and 150 people come in to the shop and buy $1,000 worth of coffee, my earned media value is $1,000.

If 5 million people see it on You Tube and see the logo of my coffee shop in the window behind the guy and never walk in the door, the value of those impressions is as much as the sales generated by the non-existent customers, zero.

Stop measuring and valuing impressions that don't generate business impact. Start changing the conversation and creating marketing jargon that C-level leaders and Business Owners care about.

Chapter 8
If a Brand Tweets in the forest and no one sees it, did it make an impact?

We spoke earlier about the overwhelming quantity of content on social media. The real enemies of content marketing effectiveness are apathy, lack of trust and context as described in the Axis of Evil. But there is another thing that we do as marketers which triggers organic reach suicide and it's super common and generally celebrated.

The reality today is that only 1% of fans of a brand engage directly with the brand on Facebook.

98% of the fans that like a brand on its Facebook fan page will never return to that page again.

They simply liked the page as the required opt in to gain access to some discount, freebie, or special offer. They are not engaged, and they do not see what's happening on your page.

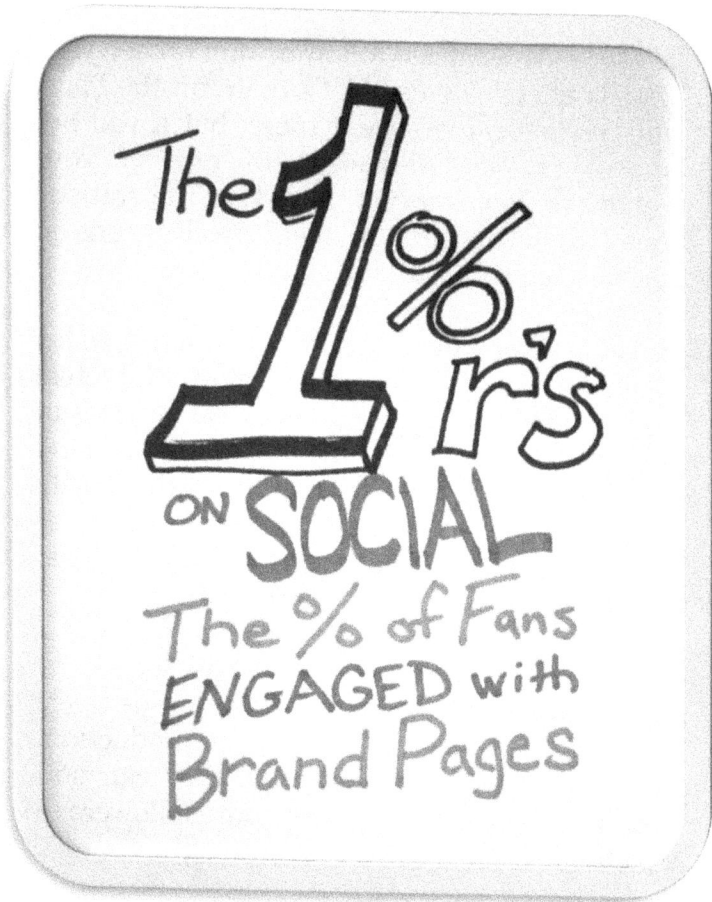

The 1%'rs ON SOCIAL
The % of Fans ENGAGED with Brand Pages

I expect that you already understand the way that Facebook Edgerank functions when you publish new content without paying to promote the post. Just in case, I will explain. Facebook shows a new post to a test group of **between 1 and 3% of your most engaged followers**, and based upon their interaction or engagement, they determine whether or not to show that content to any additional people.

If your most engaged followers don't react, your post is effectively dead. It's there on the Page, and anyone can go see it there, but it will not appear in the Newsfeeds of any of your Followers, it just won't. Based on the statistic that 98% never return, that effectively means no one will see your activity.

A common practice in content marketing today is to aim for entertainment value. Marketers know one reason people go to social media is to be entertained. Some brands spend considerable time and money to create extensive and highly engaging entertainment pieces. The hope is that such pieces will appeal to a broad range of people and go viral.

If they are successful, the post is shared by people already on their page and by a wide range of others who subsequently see the production. When a brand does this 'well' they can add hundreds, thousands, or even more followers to a page.

If we measure by illusory vanity metrics, this is certainly very exciting and would be marked as a success. If we understand that those vanity metrics are meaningless, we realize that the production is likely not a success and could actually be very damaging to your future business.

Unless the piece is carefully crafted to appeal to your ideal brand audience, adding a multitude of new Followers who are not your ideal target makes it impossible to use your Fan Page to directly reach the fans and followers that you want to engage with and have worked so hard to build.

If you recognize this activity, you have a sinking feeling in your stomach right now as you realize that you invested significant budget and creative power to build a highly entertaining but useless marketing stunt on your social network.

Since you're not in the business of unicorns crapping rainbow ice cream cones, your massive investment may have been fun but also contributed to Reach Suicide.

Reach Suicide = Intentionally adding huge numbers of Non-Prospects to your Fan Page thereby negatively leveraging the Facebook Edgerank function.

This ensures that you have no chance of reaching your real target audience going forward without unsustainable unicorn content that likely doesn't support your key business objectives anyway. We've all heard stories about companies and business leaders winning innovation awards on their way to bankruptcy.

This Unicorn Content/Reach Suicide is the Digital Economy equivalent of innovation awards for bankrupt companies. Don't do it.

We will speak later in this book about positive examples such as the very timely Candace

Payne, Chewbacca mask, Live Facebook video event. This is an extraordinary event, viral event, featuring user generated content that produces a result that a brand cannot possibly replicate but only dream of and does not commit reach suicide in the process.

And don't Buy Followers. Just Don't.

Chapter 9
Advocates vs. Influencers

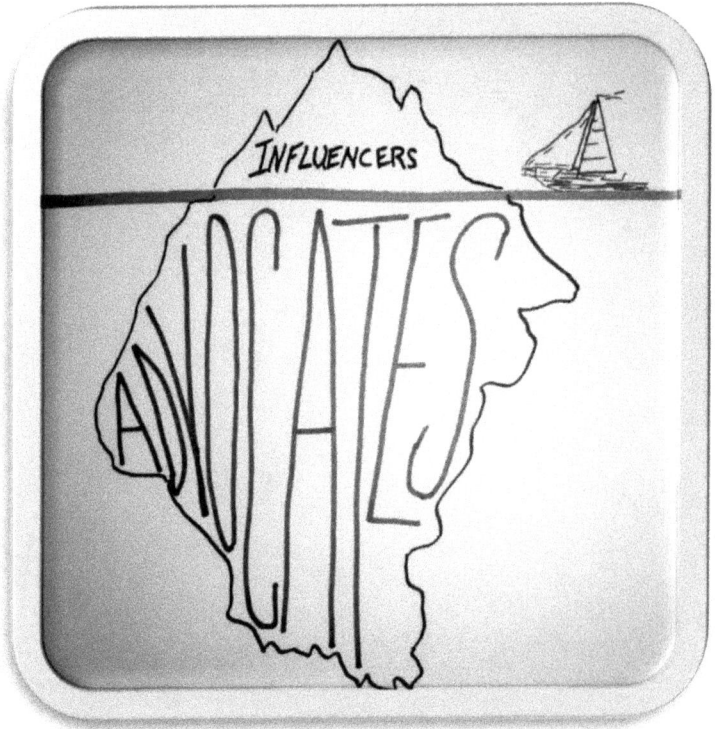

In ecommerce and advertising, a **Brand Advocate** is a person or customer who talks favorably about a **brand** or product, and then passes on positive word-of-mouth (SWOM) messages about the **brand** to other people.

Influencer marketing focuses on using key leaders to drive your brand's message to the larger market.

Rather than **marketing** directly to a large group of consumers, you instead hire / pay **Influencers** to get out the word for you.

There is no hotter topic in marketing today than Influencer marketing. A report by Thompson said that 59% of marketers are planning to increase their Influencer marketing budget in 2016.

84% of marketing professionals worldwide are planning to start Influencer marketing this year and those who are already doing so were planning to prioritize and budget even more.

The reasons for this are clear. The vanity metrics and earned media value derived from Influencer marketing programs is significantly higher and more effective than paid media. No question.

More than 90% of consumers are more likely to trust an Influencer, even if they know they are paid by the company they are representing. For these reasons, Influencer marketing is a hot topic and can be an effective marketing strategy which could demand a portion of your Marketing focus.

Earlier I demanded that we stop using military language. This will be my last use of a military term right here.

Influencers are mercenaries. Their opinion is for sale and everybody knows it. While influencer marketing is all the buzz at the moment, the reality is that Influencer marketing effectiveness peaked in 2014 and is now in rapid decline, collapsing under the weight of its own inconsistencies and spiraling costs.

While still infinitely more valuable for the company than is standard paid advertising content, Influencer marketing is only 1/5 as effective as genuine Advocate activity. The critical difference between the positions is that an Advocate is not for sale.

Advocacy is earned by customer experience and relationship to a brand.

That is hard earned and genuine value given, which will produce genuine value in return. The reason is simple. There is nothing more valuable than a trusted recommendation from a knowledgeable Advocate.

When ranked for Trust, advertising ranks at 1%, trust report Influencers at a 19% trust, and friends and family at a 97% trust (*everybody has that crazy uncle right? So I suppose 100% is unachievable and perhaps 0% is also, but we're close on both ends*).

People are turned off when they discover their referring Influencer is actually being paid. They have always known that celebrities were paid for endorsements, but for a long while people were genuinely naïve regarding the fact that many of their favorite social network celebrities were paid endorsers of the products that they showed.

A steady wave of accidental exposures has led to increasing distrust and skepticism regarding product placement among popular social celebrities or Influencers.

In recent months, the pressure has begun to mount publicly in the face of accidental exposures of the incongruence of many Influencers and among astute marketers that have been paying these influencers significant cash. This has been quietly discussed for some time, but now is reaching media discussions such as the digiday.com article recently titled "Confessions of a social media executive on Influencer marketing." My personal favorite is the new article by Andy Cush in Gawker titled "The Influencer economy is collapsing under the weight of its own contradictions".

Andy also says: "Influencers are going to start disappearing. Brands are going to start realizing that the number of followers you have doesn't mean shit. Just because photos look good and have 200,000 followers means nothing. You can't rely on content creators all day long. For the Influencers, we live and die by these platforms today. Rest in peace, Influencers. Long live the next thing."

When we're talking about five and six figures invested per tweet or Instagram Post, anybody with a C in their job title needs to be asking hard questions about just how much influence that marketing dollar buys? What is the ROI of this activity?

Especially with the frequency of the "I just love my new Samsung" phone "tweeted from iPhone" continuing to expose the incongruities.

I predict another significant pressure will come upon these paid Influencers as they continue to demand higher and higher payoffs for their influence. More astute marketers of the subject matter will begin to expect cross channel measurement and multiple network influence reach.

As I have written extensively in my work with the Party Principle of Social Networking Success, it is critical for someone seeking personal reach on social networks to dive very deeply into a limited topic space on a limited number of networks. Attempts to run wide in scope will largely be ineffective, sometimes even when commanding true celebrity.

As marketers demand greater cross channel influence of their Influencers, the Influencers will attempt to go wide versus deep, and in doing so they will reduce their effectiveness on all networks.

As they engage in new spaces as a new personality, their new community will quickly and easily perceive their inauthentic nature and their success on new platforms will be very difficult. This pressure will also take time, energy and creativity away from the unique art, which is the key to driving many of the Internet celebrities. Attempting to simply put the same content on multiple networks is never effective, because they each have their own unique culture and style.

As the Influencers are required to invest additional time, energy and creative resources to master multiple networks, their effectiveness will decline significantly in the primary network that got them the influence they are being paid for. Mark my words on this trend.

In a recent survey of active Brand Advocates, only 1% said that hoping to receive some pay or reward was the primary reason for recommending a product or service. Among genuine Brand Advocates, their highest interest is to help others and their highest social capital is earned by their in-depth knowledge and expertise which is both valuable and appreciated by their peer network.

The key takeaway is that genuine Brand Advocates bring a true sustainability to marketing.

Advocates can be your best salespeople. More than that, an effective brand advocacy program can and perhaps should be your primary marketing function.

As would be expected, some Advocates are clearly more valuable than others. There are natural Brand Advocates, sometimes called Super Advocates. Some of these have great reach that extends into the thousands and even millions who can be reached with a trusted, relevant, message.

As you and other astute marketers understand the shifts in Influencer marketing, I'm confident you will see three clear focuses being elevated as its replacement. These are pure advocacy plays:

Trend #1: An increased focus in employee advocacy. 31% of high-growth firms now have formal employee advocacy programs in place. In the future we will see brands using employees as Advocates in order to seek the trust and loyalty of their customers. Employees will be encouraged to share content, promote brand messages and target key accounts. These employees will become brand ambassadors and will also link to influential subject matter experts. This will help to deliver and amplify the brand message to the right people.

The real key here is to shift the focus of present employee and brand advocacy engagements. To elevate them, not in an expansion of how many employees are participating, but in what level of employees are participating. We will discuss this again as we get into more detail on the best practices of a successful brand advocacy program.

As you better understand this blending of marketing, customer service and customer experience, you will understand that the real opportunity here is to get higher level employees and subject matter experts in your company engaged on the company side of an effective Brand Advocate community. This could involve the employees sharing branded content on their own personal profiles, as dictated by an employee advocacy strategy.

As an employer myself, and having been an employee, I see a reason for full discussion about the appropriateness of requiring a corporate employee to share branded content on their own personal social profile. However, I see no ethical conflict of any kind, and massive value to be gained by having all levels of employees actively engaged in a successful consumer brand advocacy platform.

Trend #2: We will see a rapid rise of tier 2 or tier 3 Influencers. These are low-priced Influencers of small school subject matter expertise, and without the celebrity following that we associate with Influencers today. But they could easily have a significant influence role in a narrow scope, particularly in specific lifestyle brands and specialized B2B markets.

Ultimately, this will be fleeting and likely ripe with incongruence and lacking authenticity as the innate mercenary activity of a paid Influencer always will be. That does not mean it won't be popular or that it will not prove to be more effective than paid advertising alone.

Trend #3: With the rapid rise of professional and even higher paid Influencers, there will be an increase in demand for authentic, natural Advocates who act like the Influencers of today.

We will see greater collaboration of organic Advocate outreach.

Unpaid but niche influential Advocates will become the most successful outreach strategy on Social Media as subject matter experts and celebrity social media icons become more difficult to work with, more expensive to engage, and less effective in their results. Brand advocacy is not fleeting.

This is a renewable business asset which can last for many, many years. Think about your own personal or family consumer behaviors. Many of your brands are probably your mom's brands and perhaps her mom's brands. If something works, we often tend to take too few risks and stay loyal to brands we trust for a lifetime.

As an example, I have a close, personal relationship with Jiff Peanut Butter and I'm sure you have such connections as well.

Chapter 10
#CX is inseparable from #SMM

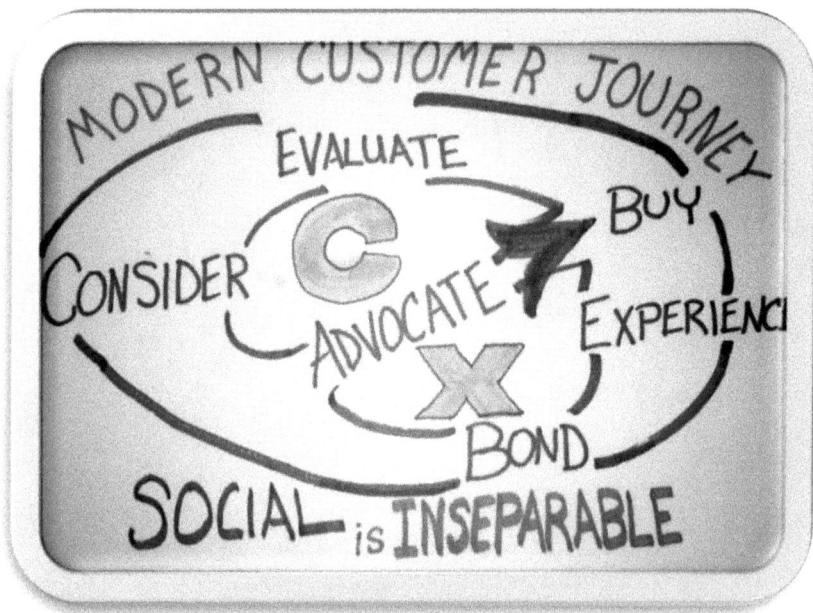

Forgive me for testing your hash tag savvy, but today your Customer Experience is inseparable from your Social Media Marketing presence. Any idea of having a sustainably profitable and growing company based upon a single product or transaction today is utterly foolish and not investable.

Companies must break down any barrier between sales, marketing and customer service. They must recognize that their customer experience and their marketing are walking hand-in-hand from this day forward.

When you map out a customer decision journey for a modern empowered customer today, you must understand the active presence of social media interaction at the Evaluate stage, the Buying stage, the Experience stage, the Bonding to the brand stage and then hopefully, for your sake, the Loyalty and Advocacy stages. That consumer journey has social interaction written all over it.

There are four types of clients that will engage actively with the company on social reporting concerning their engagement with your brand. They are the Advocates, the Followers, the loyal Fans, and the broader Community.

The Community is engaged to learn. They want to be involved from a technical knowledge standpoint and they are ripe for engagement as Advocates.

The loyal Fans are customers primarily due to habit or convenience. They tend not to be overly excited about the engagement other than for convenience, but it is possible that as many as half of these folks could be engaged with and motivated, or seeded, to become actively involved in your marketing.

Followers mostly tend to be just that. They tend to be followers and they are along on social media primarily to get discounts and be aware of what's happening at a surface level.

Brand Advocates are those who are loyal, who have learned about your products and often your company. They want connection with the brand and technical knowledge at advanced levels. They have an innate desire to help others and to connect others with a valuable purchase experience with your brand.

Clients who become Brand Advocates see themselves as part of the brand and as part of your company. They are far more valuable than normal customers. Brand Advocates are in fact worth at least 5X your average customer.

Brand Advocates will typically spend twice as much as your average customer and will typically refer three times or more than the value of their own consumption.

The value of Brand Advocates will increase proportionally with the value of your average sale. A five times return brought by a big fan of Ramen will not be as valuable as a five times return brought by a big fan of BMW. Sorry Ramen, that just is what it is.

Loyalty does not necessarily mean advocacy, nor does frequency necessarily mean advocacy. The trigger for Advocate activity is the genuine desire to help others. Brand Advocates tend to be habitual recommenders, on average recommending products and services 26 times per year.

Consumers want brands to use social media as intended, as a two-way human, communication channel. Unfortunately, and to their own detriment, the vast majority of businesses are not matching those desires.

According to a recent study by Sprout Social, on average, businesses send out 23 promotional messages for every one direct consumer response. 23 to 1!

Social media is now the number one way that consumers interact with businesses.

Sprout Social's research reveals that 35% of consumers turn to social media first when they have a problem. The article said that by and large, people aren't asking for the world, they simply want to hear back when they reach out to the brand.

In our experience, a midsized consumer brand ideal client of Smync will receive nearly 1000 messages per quarter which could elicit a response. This is up nearly 20% from just one year ago. Unfortunately, nearly 90% of those messages that merit a response are ignored by most companies. The messages that are answered are often not done in a timely fashion. The average time it takes a brand to respond to social media message is 10 hours!

Most consumers expect a response in less than half a day, at worst. 75% of people say they are more likely to become Brand Advocates if they receive a meaningful connection on Social

Media and nearly half surveyed said they are more likely to purchase because of an interaction from a brand on Social Media.

There are two sides to this discussion. On one hand you have in the survey from Sprout Social which reports a problem and is really seeking a customer service response. A response to a problem should be nonnegotiable at this point from a customer service point of view.

It is completely unacceptable for 90% of properly directed social messages to be ignored by brands. But the other side of this discussion is the massive missed opportunity of a simple Reply/Like/Mention to a current client or prospect on social media.

When you have the keys to the social account of a lifestyle brand with which your clients and prospects identify passionately and personally, you have an amazing opportunity to build a connection which could potentially change the lifetime value of that customer by a large multiplier.

This is as simple to understand as an experience that I'm certain you have had when you are in a private, local restaurant and the owner of the restaurant stops by your table to greet you and say hello. That is a memorable experience which builds brand loyalty immediately and for the long haul. A rock star moment!

And that same feeling can be easily replicated by a lifestyle consumer brand. Why would you settle for just hitting the like button when, if you actually take a moment to engage, you can not only make someone's day, you can make an impression, a connection at an emotional level, to that brand that would be difficult for any competitor to break?

Chapter 11
Owning vs. Renting your Own Customer

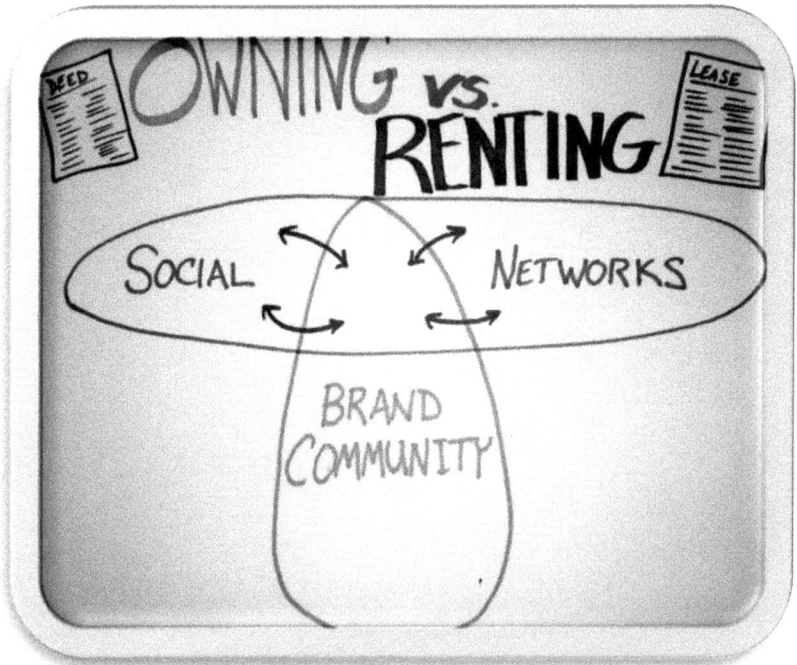

As we get into the nuts and bolts of running a successful brand advocacy marketing system for your company, we will discuss the creation and development of community.

The functionality of the community aspects of our company's brand advocacy platform is a critical aspect of that success. Along with that, we need to recognize the requirement for a swing in online communities. It is no longer enough to rely on social networks for the community infrastructure.

Smart companies are combining their own, owned communities with the network's rented communities into a 'one plus one equals three' scenario which Jay Baer calls Omni Social.

Originally, social media consisted exclusively of brand owned communities like the Harley Owners Group, built, organized and facilitated by Harley-Davidson. This preceded the existence of the new digital landlords which is a pretty good description of what the social networks have become.

With the unprecedented growth of the major social networks, most businesses have taken an 'if you can't beat them, join them' approach with online communities. Consequently, they allow those communities to be hosted entirely on the social networks.

As we discussed in the Axis of Evil, that introduces a number of grave concerns when the network then controls our list of customers, and charges us for access to them, while at the same time allowing them to be an open source list which is available for anyone to see, including your competitors.

I'll give you a minute if you just shouted 'Oh, Crap!' You are not alone. And that reality has just never fully dawned on most Social Media using businesses.

As a result, brands, probably including yours, are now renting access to your customers and to their attention. Plus, the landlord is a public company which has a responsibility to maximize its own revenue. They are free to gather hordes of data about your customers that they don't share with you, use for their own interests, make available to competitors and charge you any way they see fit to maintain contact. If that hasn't scared you yet, you're not listening.

The decline of free organic reach and the requirement to pay to gain access to your own community are just the opening salvo in what will be a protracted and expensive arrangement. This makes social communities more costly and less efficient for your brand, but from where you are today and where your customers are, it's not realistic to just walk away. For now, whether you want to play their way or not, you must. At least until you make other arrangements.

One thing you should do right now is leverage the Smync Advocate community platform to create a hybrid structure in which you own your Advocate Community and have complete control over that community.

Advocates can freely be found on, come in from and share back to any of the popular social networks. This cross network connection is powerful by itself. We'll talk more about this, but I can tell you that it is infinitely easier to manage than multiple network-based communities, each with their own customs and sets of rules.

It allows you complete oversight control and measurement of your most valuable customers which, much as you may like Mr. Zuckerberg, should not be left to him and his shareholders to control.

This will require you to do six key things:

1) Maintain an organizational commitment to building and managing your new community.
2) Committing to rich two-way communication and valuable online experience that your customers enjoy on the social networks and will expect of you.
3) Committing to creating for yourself or leveraging a subscription to the Smync community platform.
4) Managing and maintaining the flow of those customer relationships between the networks in your own community, while avoiding duplication and confusion both for yourself and your customers.
5) Communicate both internally and externally so you will satisfy the population and function of the internal and external communities.
6) Most of all, you must demonstrate a commitment to customer empowerment that grants significant control of the environment, dialogue and story inside

your own brand's community. This enables the community to be as much theirs as it is yours.

To be successful and enjoyable for your clients they must be able to interact with you on a multilevel basis and also with one another. This can in no way be a one-way street.

The benefits of the triangle shaped community will be difficult for you to fathom right now, but perhaps the vision of the creative power of what is now called crowd sourcing will give you an idea of the inherent strength of an empowered Advocate community.

Just imagine the amazing windfall of User Generated Content to fuel your content marketing monster, while simultaneously teaching your company more about your client's emotional connection to your brand than you ever could from any survey, focus group or other activity that you have likely used in the past.

Remember that **what others say about you is infinitely more powerful than anything that you can say about yourself**. Effective social brand advocacy puts that power on a whole new level. This can spell nothing but progress, prominence and cash for your company's future.

Chapter 12
The What, Why and How of Social Brand Advocacy

Less than 20% of consumer brands have a formal advocacy program in place for their company today. At the same time, according to CMO.com, roughly 90% of corporate marketing leaders are establishing such a program as a number one priority for the year 2016.

No doubt many will read this book and want to immediately jump in with both feet. That's cool, but before we dive in, let's have a look at the steps to appropriately and effectively launch a social brand advocacy program.

There are three steps that are key and standard in establishing your advocacy launch. These three steps should only take an aggressive company about one month.

1) Create your advocacy plan.

2) Prepare your internal readiness. Organize how you will interact with your Advocates and who will manage the system.

3) Prepare your technology readiness. Prepare your platform and the branding of your Advocate community with your Tech team, Social Media team and Marketing.

Social Brand Advocacy Plan

Your Social Advocacy Program Plan is simply a roadmap, a guide to reaching your desired destination. Just like any other marketing plan, an Advocate marketing plan has three major components:
1) Where you are now
2) Where you want to go
3) How you plan to get there

Your outline will contain an analysis of your present situation, your goals for the program, your action plan, and the calendar of steps, including who is responsible to complete each of those steps.

A cautionary tale here; the most talked about brands, with the most successful brand advocacy programs, are those that create a very special, intimate, lasting connection with their best clients. Therefore, as you begin to evaluate your advocacy plan and talk amongst your team, it is very important that you understand that your goal is creating a long-term love and not a short-term buzz.

If you're seeking a quick pop, there are a lot of ways for a marketing department to create some publicity, get some strategic endorsements and create some.

Creating a successful, functional, network of Advocates, (customers who support your brand for the long haul and will publicly talk about it with those they care about) will take more effort to build, but will guarantee a long-term close commitment from those fans.

The best programs empower regular people to become super fans and more importantly, virtual partners of your business. They are the ones who will go the extra mile to get the word out about the great experience they have working with your company.

You must seek Community first, and recognize that a strong sense of belonging and connection will be critical for long-term success.

The outline of your Advocate marketing plan will at minimum, include:

    1) your present situation with regard to customer connection points;

    2) agreed-upon goals for your advocacy program;

    3) an action plan, including all of the key figures in your company that must be committed to the goals and involved in the plan;

    4) the calendar of steps in the plan

The evaluation of your present status must be an honest one and should begin with understanding the state of word-of-mouth communication about your company on social networks today. This can be a very challenging data analysis task if you don't have the right digital tools.

It is important to evaluate all social network's review sites and forums to understand the current state of your online reputation. The tools on our Smync social relationship platform will save you a ton of time here.

An analysis of the potential for amplified social talk on your behalf is next. The quick analysis of Smync Score will provide you some initial feedback for those already engaged. Depending on the size of your customer database and the reach of your company, you may require some additional analysis, including survey tools, to determine your net promoter score and the potential for your company.

Conversations with your sales department can provide valuable insights on the status of word-of-mouth referrals today. Engaging salespeople and customer service staff will likely be a great source of information.

This will help you quickly identify likely prospects for super Advocate status and prevent missed opportunities for folks who are not yet appearing in social data.

Depending on your product line, you will likely want to analyze a variety of product review sites and company review sites, if you haven't been doing that already. From all this you can compile the data for an analysis that you can bring together for the Advocate development team.

You will also want to make sure that you have baseline data of your sales and market share today.

Brand Advocacy Goals

What will be the focus of your advocacy efforts? What are the goals your senior leadership is seeking?

1) Increased advocacy activity regarding specific aspects of the brand; or specific product lines; or services; or categories?
2) Increased advocacy in a particular geographic segment, or on a particular social network?
3) Do you seek to combat existing negativity? Or to amplify existing positivity?
4) Is the real driver of the program to support a new launch or some special initiative?

Brand advocacy goals are generally designed to support the entire company's marketing goals. But in some cases, your company may be seeking to expand a certain market or grab market share in a certain space. Perhaps you are preparing for a special endeavor and if so, you need to be aware of that in designing your goals.

There are existing texts on brand advocacy with general and specific detailed descriptions of each step required to create a brand advocacy program from scratch.

I will not repeat the detail of those books for two reasons:
1) my focus and our company focus is specifically on social media enabled Brand Advocate interaction;
2) because of the build out of Smync and the success of those tools we have already deployed, I have complete confidence in our ability to execute a well-designed plan that no longer needs to be as detailed as was required at the time that the book on brand advocacy was written.

But there is a specific section referenced in the book on brand advocacy and used by market leader Zuberance to create a marketing program from scratch that is relevant and will be helpful to you. It uses the acronym POST to organize the information required for the creation of a brand advocacy program. (POST as a method comes from Charlene Lee and Josh Berghoff, co-authors of the well-known book GROUNDSWELL.)

P is for People: Based upon your expectations or the objectives of the program, what type of clients do you presume are going to be the ones to engage with you in social community and be willing to share on your behalf? As with all effective marketing, having a clear picture of this perfect, ideal client avatar is essential to all marketing work.

The rapid analysis that comes to you via the Smync score on the Smync platform will provide you with great insight and quickly help you determine if the people who are already representing your brand online are doing so in the way you desire.

You can also determine if they are a good match for the avatar expectation in your collective minds. This insight will be very helpful in deciding how quickly you will be able to engage this audience in your private community.

O is for Objectives: goals are general, objectives are precise. What are your advocacy objectives? They should be clear and measurable.

S is for Strategy: your advocacy strategy describes how you will achieve your objectives. If your goal is to energize 1000 brand ambassadors around the launch of the new product line, how do you expect you will do this?

T is for Technology: the technology and resources available to you inside the Smync proprietary platform will undoubtedly blow your mind and exceed your expectations. So instead, **this T stands for Tactics**: when you strategize how to engage your loyal fans in community, think creatively.

Imagine the types of content you will share with them, the types of offers you would like them to share on your behalf or the specific game of vacation elements which might be inspiring to your unique audience and motivate them to further engagement.

Imagine how you might connect the members of your community with key members of your team or key Influencers, celebrities and athletes that you may be engaged with, and who would be exciting to your hard-core fans.

This is a good time to begin your mindset shift by using the likely candidates for advocacy that you have already identified in this initial research. Invite them into a mastermind or focus group and engage them in the creation of the community before it is ever launched.

Companies often forget to ask for feedback and participation from their most important audience. You are moving into a new level of connection here, so why not begin by breaking down the barriers between your company and your customers? Involving a small number of key customers at this time will likely produce a crop of super Advocates who will engage in your community immediately and boldly.

This will help set the tone for the level of participation you desire as others are brought into the mix. Involving key clients as a customer advisory board in designing your Advocate program will also help to adjust your company culture to the idea of a more direct and open communication with customers.

It will also give you an idea of the level of exclusivity and connection your customers want and need to be motivated. Additionally, it will give you the first information to demonstrate the long-term benefits for anyone inside the company who is not yet onboard with the strategy.

Think belonging and community first. Belonging will be critical and this masterminding activity will help greatly to set the tone for your Brand Advocate community. Many of us are familiar with the concept of 'jab, jab, right hook' that is explored in a book written by well-known social expert Gary Vaynerchuk.

Inside your Advocate community your engagement will likely be more gradual. Something like jab, jab, jab, jab, jab, jab, right hook: product insights, knowledge development and education, fun, more insights, special privileges, and even more fun. It will be critical for you to emphasize fun in order to develop real engagement. People are expecting an experience which is equal to or better than that which they enjoy on the social network that they have come from into your community.

At the point where you have delivered this experience, you will have earned the right to ask them to share on your behalf, and when you do the results will be spectacular.

Lastly, in determining your objectives, no doubt you'll want to determine your measurable goals or KPIs for the program. When you do so, do not allow the virus of vanity metrics to creep into it.

The number of Advocates, numbers of likes, impressions, and all those useless numbers will not help your company create an effective social brand advocacy platform. Instead, identify what would be homerun instances of advocacy and create a plan to trigger and track them.

Instances of advocacy trump numbers of Advocates.

Begin to think about the amplification opportunities of your Brand Advocates. Once you've created an amazing community of Advocates and have them engaged, the awesomeness will extend well beyond simple social sharing. Existing influential Brand Advocates work 24/7 around the globe to amplify brands' messaging, provide endorsements and referrals, build brand communities online that share and amplify the reach of paid advertising campaigns.

They may even wear their status as a Brand Advocate for your company as a digital badge of honor, wherever they may go online.

The most important thing to remember for success with our social brand advocacy program is that we are in it for the long haul. We are playing the long game. Plan ahead to keep your members engaged and interested. Plan to go the extra mile for your engaged Advocates. Show them that you truly care, never take them for granted, and provide them with a true VIP experience.

When the time comes that you need their support, they will be there for you. Every time, and for the long term. Long-term commitment from your company will build long-term loyalty from your customers.

So again I say: drop the military lingo. Stop targeting campaigns and start building community. Create a movement that pulls in customers and makes them happy to stay!

The influence of your community will be exponential and far more influential than any big-budget Influencer that you might rent!

Chapter 13
Measuring Return on Advocacy

The Smync platform measures many things including an exclusive and patent pending return on advocacy tracking metric. (ROA)

Other simple measures that we will watch as we get each client off the ground and make things run well for them include:

1) Which members of the community are sharing the most or are the most engaged?
2) Which members are effectively creating user generated content?
3) Which members are popular and influential within your own community? The Smync score will easily and clearly show which Advocates have the highest external connection to the brand. We will also identify those that are playing the game as we want them to, and who are the most influential inside your company and inside your community.
4) What content is being shared most?
5) Uniquely, inside this community, you'll be able to follow up each question with the question of why? You gain insights as never before on the effectiveness of your marketing strategy and investments. At the same time, you will gain unprecedented, valuable input for

product development and general corporate strategies.

6) On which networks, forums, blogs, etc., is your content being shared? The sharing practices of your best clients from the inside platform will provide you tremendous insight into the most influential external platforms. You will have the ability to see this activity through the exclusive Smync ROA system and make significant improvements to your marketing strategy, content strategy and paid media strategy.

7) What is the program's overall impact on purchase actions? You will be able to track the impact of your content, your Advocates content, and conversation stations and sharing as driven by the Advocate community. And, to the extent that your business ties to online e-commerce action, we will have a very clear path of social activity to sales.

This is why Smync so confidently claims the ability not only to facilitate brand advocacy, but to measurably prove the direct financial benefit of that social activity. This is the proprietary Return on Advocacy metric.

The return on advocacy tracking tool is not only proprietary to Smync but is also proprietary in its function on Social Media, its ability to pass and maintain its track from social into digital and back to social and then to e-commerce action.

There are also existing relationships in place which are still under development. These link return on advocacy to direct retail action at a store level. There are currently two strategic partnerships in play which will bring this to life in ways that, until now, only existed in wild marketing fantasy.

In a recent engagement by Microsoft in their small business space, their Advocates strategically shared a special offer with an 80% response rate. This generated a 66% growth in their small business services in just one month. Small business decision-makers are extraordinarily difficult to reach for marketers, but not for other small business decision-makers who have something of value to share with their colleagues. This is Social Brand Advocacy at its best, driving rapid business growth at nearly no cost!

Chapter 14
Exponential Influence

Every day I drive past one of the weekend homes of each of the founding families of the Amway Corporation. Both impressive lakefront properties are roughly 50,000 square feet, and, I assure you, fit for the lifestyle of a billionaire. And yes, I said one of the weekend homes. With this frequent exposure so fresh, perhaps the effect of the exponential influence of direct word-of-mouth marketing is more real for me than for others who may not see it so close-up and so often.

There are two sides of the immediate impact of exponential social word-of-mouth that I want you to understand and get really excited about.

If we return to the Jeff Bezos quote about our brand being what the marketplace says it is, then we have to recognize the critical influence of our online reputation today. That reputation is driven by individuals publishing reviews, recommendations, posts, and sharing comments. By effectively harnessing an outstanding group of social Brand Advocates, you will be able to positively impact your corporate brand reputation as both an offensive and a defensive strategy.

As you establish community and increase loyalty and connections inside your Advocate community, one of the things that happens is that you give clients visibility and easy access to opportunities for them to review and recommend your products and services in the online context. With good relationships that are firmly established, it is an easy ask to have your Brand Advocates share their 'brand story,' or their experience with your brand, doing this in spaces that are highly influential and will have long-lasting beneficial impact.

On the other side, what we have already frequently seen, and you will soon see for yourself, is that when either a brand 'social media emergency' erupts due to some embarrassing situation, (which will certainly occur in the life of every brand,) or with instances of noisy negative reviews online, your Advocates will come to your rescue with authentic and exponential influence.

Your Advocates will counteract any negative social reputation consequences for your brand with a speed and exponential power you've likely never seen before.

Chapter 15
Sharable is the new Viral

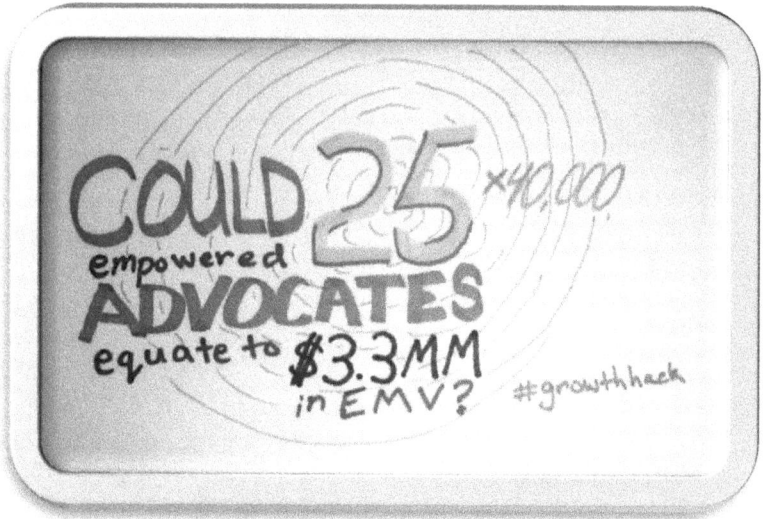

COULD 25 ×40,000
empowered
ADVOCATES
equate to $3.3MM
in EMV?   #growthhack

Now that you understand the cautionary tale of artificially inflated social following that destroys organic reach, (discussed in the Axis of Evil,) you will recognize that the new and more valuable viral is 'shareable.'

Unfortunately, viral as it is normally measured can be destructive and is certainly unsustainable. Shareable information insights, experiences and stories via your Advocates and about your brand create a sharing that is far more valuable than what we think of today as 'viral content.'

In his book BRAINS ON FIRE, Spike Jones explores that one empowered Advocate sharing on Facebook can ripple out to eight generations of Connections. This equates to 40,000 Impressions delivered in context and with a trust that cannot be gained through paid impressions.

I like to extrapolate that statistic. It takes 25 Advocates (a number that even a small startup could achieve) to create 1 Million Impressions. In context, trusted impressions.

By using one definition of Earned Media Value combined with a standard Facebook Advertising menu, we can estimate that reach as commanding the equivalent of $3,300,000 in Advertising spend. Free.

Well-known Facebook Marketing thought-leader Mari Smith has said that the key factors for highest shareability in content are things that make people either Laugh, Cry, or say 'Ahh." This is an extremely tall story telling order for a Brand. Logos don't make people laugh, cry, or say 'ahh,' people do.

In order to enjoy the benefits of social connections and sharing, brands must boldly embrace the co-creation process of brand advocacy. You must get out from behind your logo and build a community that can both tell a better story and tell your story better, while they connect in a real relationship. People want to belong and gather and tell fun stories. That's where success is found today.

Chapter 16
How to Execute Social Advocacy Strategy

As with nearly every strategic endeavor, there is an easy way and a hard way. The easy way is fast, gets big results and sets you apart.

To Create and Execute your Advocacy Plan:
1) Make your planning simple. Your plan doesn't need to be more than a deck of 10 slides. Cutting-edge marketers who want quick results will leverage the Smync platform. This will make planning easy and save time.

2) Keep your planning process short. No more than a month to create a whole launch. The marketplace and the networks change so fast today that if you think too long things will be different before you are ready to get rolling.

3) Focus. Don't try to change your whole corporate culture or marketing plan with one social advocacy launch. Focus on a specific segment of your customer base, a single product segment and a single outcome. The

narrower your focus, the easier to produce measurable results.

4) Stay SMART with your objectives. (Specific, Measurable, Achievable, Realistic, Time bound)

5) Get buy-in from key stakeholders in your company. An effective advocacy program requires the involvement of multiple levels inside the company. Make sure you have all of the coordination you need with sales, digital marketing, CRM management, etc.

Even though we are focused on engaging social brand advocacy, Advocate community coordination and communication will seem risky or scary to some companies. It is easy to nod your head and agree with the statement that your customers control your brand identity in your marketplace. It is another thing to truly accept that, including deeply understanding the intimate communication with your customers that will occur inside the community.

This acceptance means you will release significant control over the detail and the tone of your brand message to a group of customers and external stakeholders. Perhaps you are the person in your company to drive this, but if you are not, you must seek a high level advocacy champion within your organization.

The person must be a marketing innovator and risk taker. Someone who has the vision to see how advocacy will benefit the company in the long term and be willing and able to curtail the naysayers and sell the program as needed in the C-suite.

Building and leveraging relationships with Advocates is not simply a tactic or a fad; it is the future of effective marketing. Because it represents substantial change, it will frighten those who fear losing control of brand messaging, despite the fact that they lost control of brand messaging years ago without realizing it. Besides you never really had that power over the marketplace anyway.

Ideally, your Advocate champion will be a C-level officer or an equity-holding leader of the company. The champion's role is crucial to your success. This person must:

1) Actively promote the value of brand advocacy and justify the organization's investment in greater social media engagement and specifically social brand advocacy.

2) Ensure you have the marketing resources for the program.

3) Navigate any potential roadblocks to communicate effectively internally and externally and execute well.
4) Help with the herding of any outside agencies that are needed to effectively execute this program.
5) Participate in all major decisions along the way.

Unless your brand advocacy program has an enthusiastic and effective champion, it will be neither effective nor long-lasting. One giant mistake is common. Because Social brand advocacy is a very Social thing, many companies initially relegate the active Management of the Program and Community to a junior level employee, one who had been previously responsible for social media engagement or content. Or it may become an additional 'hat' for a Social Media Manager.

This might seem to make sense given the scope of the project. But there are several cautionary notes. The active manager must not be overly focused on social vanity metrics or specific tactics in the scope of managing the advocacy program.

Co-creation and Community will be hard to measure by traditional KPI's.
Do not allow any outside contractor or consultant to serve as an advocacy program manager. They are not part of your equity structure either financially or in spirit.

You must match the Advocate 'altitude' in the company hierarchy with the significance of its impact on your future. This match requires the gravity, wisdom and authority of a senior executive. It is not an appropriate role for some social or tech savvy 'kid' in your Marketing group or one of your support Agencies.

The advocacy program manager/champion must have the tools and the skill set of an effective project manager and a clear vision of the connection between customer experience and social marketing. He/she must carry the horse power by title, attitude and energy to coordinate the key participants necessary to create an outstanding Advocate experience.

Only then will this produce outstanding advocacy performance and an amazing Return on Advocacy. For this reason, a long-term successful PR department is a better representation of the resources needed to create a successful brand advocacy program.

These are the corporate and human capital initiatives that I focus on in setting up this structure. Past texts on brand advocacy have included this process but focused primarily on the tactical side and the technology.

There is little need to focus much on the technical side of the technology now, because the platform created by Smync has already done the heavy lifting of all that trouble, which was an impediment to effective brand advocacy in the past. With the use of Smync, you have a turnkey advocacy system and fully integrated mobile technology platform ready to hit the ground running.

This is a massive leap forward from the past and makes any information about Brand Advocacy Management from 2015 or prior irrelevant today.

There may still be reasons why an advocacy program could fail, but our experience and your prior planning can prevent these specific fatal mistakes:

1) No champion in place, or the champion is too busy or too low in the company hierarchy. Get your champion in place, get your other C-level staff on board and ensure ongoing moral and attitude support. Since this is a considerable investment of time and resources, it is essential.

2) The plan is incomplete or unclear. Get a clear plan, clearly defined and measurable objectives and agreement from all key parties.

3) Limited access to key customer touch points. This will both limit the effectiveness and fatally damage your advocacy program. To create a VIP experience for your VIP clients you

want to have all of the customer contact points on board and committed to two-way communication.

4) Don't sabotage yourself. Other aspects of your own marketing structure, or outside agencies can sabotage the effectiveness of the Advocate program. This could come from fear of turning over creative actions and brand messaging to this new cooperative customer engagement space.

Jealousy and unfounded fear can get in the way of making your advocacy program an important part of your marketing mix.

You need to have clear permission to significantly change the status quo for all levels of customer interaction and perceived brand messaging control.

Given all that, here's why you need to jump in now, with both feet!

1) You will be late to the party. If your key competitors are not already leveraging social brand advocacy, they soon will be. Brand advocacy was the number one priority reported by CMO's worldwide for 2016 in multiple studies. This is a huge opportunity, don't be left behind.

2) Technology can be a barrier or an enabler. This is a major customer centric system and technology and communication must be fully functional for it to be successful. The company cannot hide behind technology and walls. Your company must embrace your customers as a Business Asset.

3) The sooner you invest into customer relationships in brand advocacy the bigger your ROA will be. If your company has 1 million customers worldwide, you may have half a million potential Brand Advocates. If the prospect of hundreds or thousands or hundreds of thousands of your customers actively engaged with your company and sharing your message doesn't get you excited I don't know what could!

Customer social brand advocacy is not about the money, it's about connections. Simply stated, you have the opportunity to have the majority of your customers actively and intentionally participating in the growth and development of your company.

Remember, just a small increase in Brand Advocate activity on behalf of your business can double your present revenue growth rate, and do so in a highly profitable way.

Chapter 17
Finding your Brand Advocates

The new standard for measuring brand advocacy is called 'Net Promoter Score.' We've already discussed the two questions which are used as the basis for the net promoter score.

1) "Would you recommend our products and services to a friend or colleague?"
2) "Would you be willing to do so publicly?"

A further relevant question would be: "Are you active on Social Media?" Surveys to gather 'Net Promoter Score' data using these questions can be constructed and administered in many ways. Whole texts have been written about creative ways to gather that information and what to do with it.

Those scoring 9 or 10 on a scale of 1-10 are ripe to be Promoters

Those scoring 7-8 on a scale of 1-10 are considered Passive or Indifferent.

Those scoring 1-6 are considered Detractors and should be referred to Customer Service to make reparations, if possible.

Many companies have more than 50% of their customers as potential Promoters. This is AWESOME and holds MASSIVE potential. Even with a much lower percentage of customers as potential Promoters you will still have a significant number of potential promoters.
Depending on the size of your customer base, you might have a large number of them, just sitting there, waiting to be recognized and embraced as partners in developing the company they love.

Smync Score

I described the Smync 'Net Promoter Score' enough so you can understand the concept and grasp the essential concepts in terms of percentage and potential of your customers. Gathering the data for this is easy and fast now with technology. With the Smync platform you will have access to the Smync Score.

Within a few minutes of connecting your company's social profiles to the platform the software will do its magic and you will have a comprehensive list of those customers who are your champions. They will be active on social media, engaged with your brand and in many cases, already acting as Brand Advocates simply by their natural behavior and desire to help others.

The challenge of harnessing these powerhouse customers (which has been the subject of multiple books and training workshops and even whole corporate initiatives,) can now be solved by Smync with greater accuracy and in just a few moments time.

The Smync Social Relationship Manager is agnostic to the network, but keenly focused on the person that is talking about your brand. It will find each advocate based upon who they are and bring in their activity from each network, sorted not by the network but sorted as related to that person and their relationship to your brand.

This human to brand approach is unique and part of the critical mindset shift from tactics to relationships. The whole collection of these advocates can then be sorted by the Smync Score to find those who are most to least engaged with your brand on a social network. Just a short time ago, finding this information would have taken you months. Today it's minutes!

Smync has access to a fire hose of the raw data from the networks themselves. This is far different than the implied data that is shown by the networks interfaces to unsuspecting gawkers. Through the use of this advanced social listening,

Smync has made a huge challenge very, very simple. Smync Social Relationship Manager tracks the activity you would want to track:
1) Social Mentions
2) Likes and Reactions
3) Check-ins to locations
4) RT's and Shares
5) Reviews and Comments

You can even assign your own estimation of Earned Media Value for each social engagement and produce simple and clear reporting on the effectiveness of your Social Media activities in addition to the proprietary Return on Advocacy traction that I described earlier.

This is a great asset to help with justification of investment inside corporate structures as you replace traditional (fairly useless) Earned Media Value numbers and mere Impressions with proper metrics.

Once you have your Smync Score data, it is as simple as one click to invite that person to join your Advocate community.

Facilitating connections and growing relationships with your customers couldn't be simpler. You will be replacing simple 'selling to them' by setting them up to successfully sell for you.

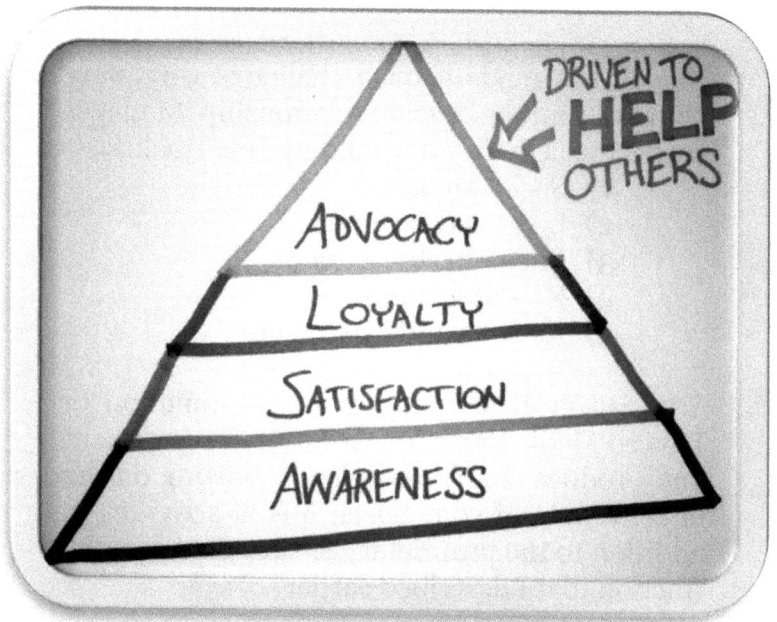

Chapter 18
What to do with your Brand Advocates

Once Advocates are engaged on your Smync managed community platform, it will be easy for you, easy for them, and fun for all to elevate social engagement.

Here are just a few of the actions which can be easily triggered from inside the platform:

1) Getting product and service reviews and testimonials on your own web properties, social networks, and on other third-party tracking and review sites such as Yelp.

2) Contests and other participation activities. People like contests and games that create competition and provide recognition. Obviously, you can engage them to spread the word about your own contests. But, more importantly, you can engage them to elevate your company in the popular Internet-based vote contests which create awards, recognition, and other publicity for your company.

   This can be very valuable and costs nothing.

3) Company and industry events. Brand Advocates will be key participants in the

events and keen to participate. You can mobilize your Advocate community to attend events, participate in your events and bring their social proof into your life environments.

There is much case study evidence to justify the value of having keyed up Brand Advocates involved in live events. It makes them feel specially connected to your brand in a unique way, and provides ample opportunity for social sharing and very cool user generated content. Do not underestimate the Advocates genuine desire to help others and to help your brand with regard to live events.

You will make them feel engaged and valuable, while serving your company growth. Make them empowered volunteers at special events. Another true VIP Experience at practically no cost to your company.

4) Your social Advocate community is an outstanding place to aid your product design and development. This is consistent with the Advocate development principle of having clients engaged with the development of the company.

Advocates want to be engaged with your product development and to participate with active feedback. In many cases super Advocates are more knowledgeable about the applications of your product than you are. Having them involved in this process will make them feel connected to you like nothing else. It will ensure their active promotion of that product or service for the lifetime of the product because they were involved in its creation. It becomes their baby too.

The Advocates' ability to participate in this way would include things like product previews and feedback for seasonal or 'next year' design changes. Using the Smync community, as an Advocate manager you are able to share content with your Advocates and also control their ability to re-share that information.

You won't need to worry about leaking things that you don't want to leak, but at the same time you can take advantage of the power of strategic leaking of new product releases into social media, which can fuel massive prelaunch demand.

5) A key driver for successful Brand Advocate activity is their desire to seek knowledge about your products and services.

Another key driver is their desire to share that knowledge with others. This adds to their social standing as a recognized expert among their friends or colleagues, and fuels the clear 'we' culture and millennial consumer desire to help others make better buying decisions.

Empowering your Advocates and fueling their hunger for knowledge gives you the opportunity to provide them with exclusive insights, training, and offers they can strategically share with their own network of friends, family or coworkers.

Case Study data has consistently shown the conversion of Brand Advocate shared offers can exceed 100%, and always is at least a multiplication of any normal market conversion rate.

Remember, the Advocates' unique understanding of the timing and context of the need for your product or service allows them to deliver your message or offer it into the right hands at exactly the right time. This explains the 'off-the-chart' conversion rates.

One significant shortcoming in measuring Brand Advocacy programs in the past is the absence of Smync's new proprietary Return on Advocacy tracking link system for information shared from the community. In the past (and with current competitor systems) only public social sharing could be tracked, and even then it could not be tracked through to point-of-purchase.

With Smync's new systems we can track the share through digital messaging. This will be the substantially preferred sharing vehicle for an Advocate desiring to deliver an offer to just the right person at just the right time.

Just think for a moment about your own behavior, if you had something to share with a friend and you knew that product today, you would not share it to a Facebook newsfeed. You would send it in a direct message, email or text. With Smync ROA, we can track this and thereby improve the measurement of the effectiveness of your brand advocacy activities.

Additionally, this tracking will allow us to uniquely determine your most effective Advocates. This will allow you to give these super advocates positive feedback about their advocacy. You could even choose to reward them in some special way.

Timely, accurate feedback regarding their effectiveness is exactly what they're seeking and exactly what they need. Remember, we're not talking about mercenaries, were talking about authentic fans organically committed to helping others.

6) This strategic social sharing can be done both on an ongoing basis and to mark

special events such as a product launch, a new location or any other specific or immediate need.

Current but less effective tools are sites like Thunderclap or Kickstarter. These are examples of socially driven pre-launch promotion. Imagine perpetually having the ability to Thunderclap new product launches.

Chapter 19
Social Proof at Scale

In the previous chapter I listed initial tactics available to engage your social networks and customers as Brand Advocates. I know my audience is sophisticated and already imagining the many things that you will do with a fired up population of connected Brand Advocates.

In addition to the previous list, I'd like to add a few other ideas on how you might use 'social proof at scale:'

1) Your new and unique ability to communicate with your client base creates the opportunity to gather customer case study data in a way you have never been able to do before. Advocates can be encouraged to document and share their own stories inside the community, particularly on special occasions.

   Such stories are significant examples of user generated content that could be pure gold! The personal stories of your best clients will become the jet fuel of the future of your marketing and company growth. This is especially true when they are shared out through your Advocate community by other Brand Advocates.

   Advocates will be excited to connect with the storyteller in this unique way and share stories with friends and family.

2) Sales References: having customers available to the sales department as testimonials powerfully illustrates social proof at scale. Advocates would be proud and excited to have their stories leveraged as specialized testimonials and prospect engagement opportunities.

   Today it's a fire drill. If a request comes from sales for an existing customer sales reference, it's a scramble at best. The data is not available, not relevant or not organized. Those responsible for managing it are fearful of engaging with customers directly. Consequently, the same few customers are repeatedly called to serve as sales references.

Without preparation, we have no idea if this is desirable or pleasant for them. Many a good customer has been worn out in service as a sales reference because of this lack of preparation and a failure to have a cadre of Advocates.

Your Smync managed Advocate Community makes this task easy, timely, and can be designed to interface directly with your CRM system to make this nearly automatic.

3) Brand Advocates can be encouraged to share and participate strategically. A first obvious thought is a Facebook post, or perhaps the strategic reference email just mentioned.

You also have the ability to direct the activity engagement and social sharing of Brand Advocates into very specific communities such as a LinkedIn group, or a specific online forum. Not only can Brand Advocates effectively share with those very specific subsets of people that you want to reach, but by becoming engaged there, and by being empowered by the company, they will serve as strategic selling tools, strategically placed customer service agents, first-line information sources and a host of other beneficial functions.

On top of all that, they will likely be excited to do this because of their specialized knowledge and the profile they receive in the Advocate community coupled with their innate desire to share and serve.
You have a free customer service department to answer questions, generate leads and referrals all day every day, all over the world.

And finally, having their presence strategically placed in those subset communities provides for immediate and active social defense in the event of negative comments or reviews within subset online communities.

There is no way your company can be aware of negative comments or reviews inside niche, subset-communities online, but your Brand Advocates can and do see these things. As empowered and excited Advocates, they can and will come immediately and overwhelmingly to your defense.

Chapter 20
Social Community
Development

COMMUNITY
ENGAGEMENT
1) EMPOWERING with KNOWLEDGE
2) AUTONOMY for CREATIVITY
3) CONNECTION to PURPOSE
CREATE A SPACE FOR THEM.
#UGC

As you leverage the Smync platform and Smync community system, you make creating and communicating with Brand Advocates and inviting potential Brand Advocates fast and easy.

It is absolutely critical to the success of your program that it be easy, fast, and fun for them to connect with you. There are a few additional critical considerations in developing your Brand Advocate community: First, do not pay them. I repeat, do not pay them.

Genuine Advocates don't want to get paid. Offering to pay them will be insulting and actively push away the people you most want to attract while at the same time actively attracting the ones you would never want to attract.

Do not pay them.

Second, do not create simple 'if-then' scenarios in which they are recognized or rewarded for specific behavior based on simple situations. This may seem like strange advice, but if you do that, the transparent and simplistic nature of your intention and relationship will be offensive. We want to attract creative, sophisticated, engaged people who will be outstanding Advocates.

If you make the environment too simple and the behavior reward structure and mechanism boring and obvious you will push away the most creative Advocates and suppress the creative energy and engagement of all of them. You do want to track them and engage them in game play scenarios where the feedback for what activities are positive and valuable is both clear and personal to each of them.

Productive people solving advanced problems work at their best when given clear feedback and allowed to use creativity to design their own path to individual and community success.

Above all, listen to them. Let them know they have been sincerely heard and are truly appreciated.

We are talking about empowering people to grow our brand on our behalf on the People's media. They can and will do so in ways that exceed your expectations and are beyond your present vision if they are empowered and encouraged to do so.

The developers of the Smync platform are creating unique and customizable gamification models for you to track performance and give feedback and do all these things at scale. I know you will be creative as you discover incredible ways to use it!

Chapter 21
Personas of Advocates and Sharers

Community works. Advocates work, but how and why?

It works because everyone is an Influencer about something. Everyone is passionate about at least one thing, if not more, and everyone who is passionate about something thinks about it all the time. It doesn't really matter what it is, but everyone is passionate about something.

Both you and I would rather go to someone we know is passionate to help us understand and learn than to a faceless brand. I trust you precisely because I know you put so much time and energy into the topic that you are excited about.

In the New York Times report about research conducted by the Customer Insight Group, they described six types of Advocates who are sharers.

These descriptions will help you to understand the different types of Advocates so you can optimize your connection with them based on how they need to share:

1) **The Altruist** – they share one very specific thing. They are helpful,

thoughtful and reliable, they use email, they're connected.

2) **The Careerist** – they share to help others in similar career paths and to potentially build Thought Leadership. They use LinkedIn, are bright and valuable and have strong business networks.

3) **The Hipster** – they are the early adopters who like to try things first and will share their insights in order to help other people who are following behind. They use public Social Sharing, are young and popular, and identify as creative and cutting edge.

4) **The Boomerang** – they ask a lot of questions and like to respond. (They can often be Trolls or at least feel like Trolls because they like controversy.) They are reactive, choose Twitter and Facebook and feel empowered on Social Media where they seek to validate and invalidate through their opinions.

5) **The Connector** – folks who like to connect people together and to share information. They use Facebook, DM's and email to share. They are thoughtful, relaxed, and creative.

6) **The Selective** – they like to share, but only privately and with a posture of secrecy. They can be highly valuable, but difficult to track. Because they share

primarily in 'dark' messaging, they are nearly impossible to track without Smync RoA tracking. They are resourceful and informative and careful with their knowledge.

These significant differences are a key reason for the new requirement of the jab, jab, jab, jab, jab, right hook culture of a successful Advocate Community. You just can't know for sure which trigger will activate which Advocate until you really get to know them. Since they are all valuable in their own way, we need to provide each of them with the trigger they need to feel empowered and inspired to action.

Appeal to their motivation to connect with each other, not just your brand.

Even so, trusting you is the key to them sharing. Keep your own content creation simple, clear and on-message so it is easy to know who to share with and easy for them to provide summaries and insight when sharing. Always stay light and fun.

You're competing with the highly engaging environments where each Advocate came from. Even treating them to a VIP experience, you have to keep it light and fun and engaging.

Remember, you must keep up with that competitive pull of their originating network and keep them okay for when you do make the big right hook ask. Always give more than you get and always have fun, because an Advocate relationship can last forever, so they are worth the time they take to build.

Chapter 22
User Generated Content and the Potential of Zero Paid Media

Many marketers, (perhaps you, and hopefully your competition,) feel some fear around the concept of user generated content. The following two minutes of video will change your mind.

This is Sam O'Hare's "Fireflies." He created this video as an empowered Advocate on behalf of his beloved Tesla cars:
Tesla - Sam O'Hare's "Fireflies"
https://www.youtube.com/watch?v=dgXvE_k YNRY

At the time of writing this book, an extraordinary event in user generated content just occurred. Everyone is talking about it. I'm talking about the miraculous moments that ordinary mom Candace Payne captured on Facebook. It is the video of her sitting in her car, laughing hysterically while trying on her new Star Wars Chewbacca mask from Kohl's.

This video hit over 136 million views in just the first week of its existence.

There's no doubt that you will see this and remember this event. It is easy to see why people love the infectious laugh and the genuine excitement shown by Candace. The video was so popular that Candace was invited to Facebook headquarters and had the opportunity to meet Chewbacca himself. She's appeared on all sorts of media including Good Morning America, and the mask itself has sold out at Kohl's and every other major retailer.

Kohl's rose to the occasion and recognized Candace. They rewarded her handsomely for her amazing piece of truly viral user generated content.

It is the most viewed event on Facebook live to date. Hasbro, the maker of the mask, and the Disney Corporation, and even the actor who plays Chewbacca have all benefited significantly from this viral social event. It is Candace's laugh, of course, that makes this video so enchanting, combined with how genuine she is in the experience.

This is a major demonstration of the fact that Advocates simply tell the story better. When a customer genuinely loves and is excited about something and recommends it, their enthusiasm is clear.

The content that results, no matter how it's communicated, is trusted and of major influence over others' purchase decisions. This was demonstrated in the record sellout of this Chewbacca mask. All as a result of an amazing piece of user generated content by natural Brand Advocate. This demonstrates clearly the results that can be driven and the speed with which they can create a huge shift.

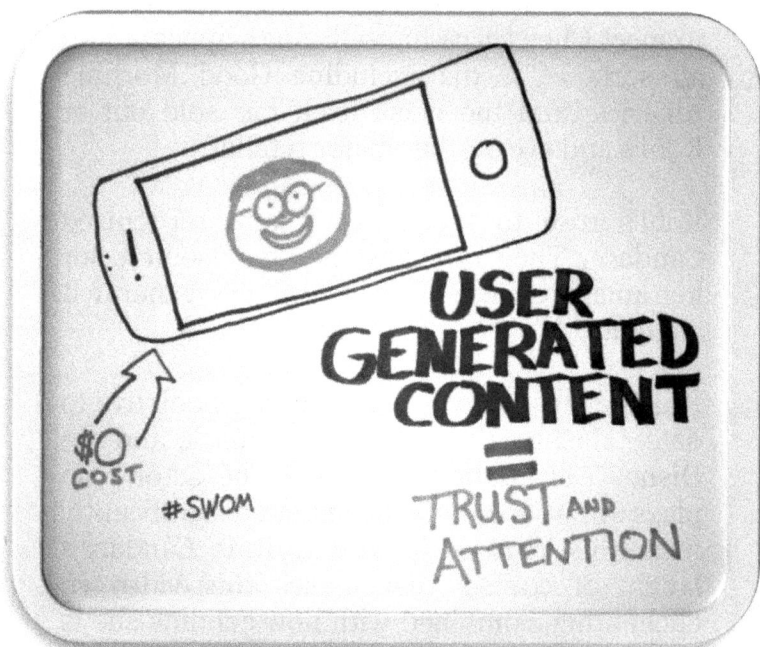

Those responsible for brand messaging are struggling with a combination of enthusiasm for, and fear of, user generated content. That fear is not hard to understand. This is a truly exceptional piece. It would be easy to be dismissive and claim that it could never be duplicated. But it will be, and soon. Coming in the most unlikely and unexpected of ways.

So, you can either sit around and claim it will never happen or figure out how to get yours next.

Companies must make peace with the fact that it is not possible to control your own brand message, if it ever was. But with social media today, it is completely different. What we can't do without serious consequence is allow our fear to prevent us from being at the helm at such an amazing moment.

You have to create the conditions so that when this type of valuable user generated content about your brand goes viral, or even just influences a few friends of the creator to take action, you reap the benefits. It is time for companies and brand managers to adopt an open mindset toward embracing their customers' content created on your behalf.

When real people, amazing technology, and positive brand message come together you have a home run, rock star, life-changing moment, that you just don't want to miss and can't possibly replicate.

The Candace Payne video highlights the power of humor in making connections as humans. Brand experiences built around happiness and joy aren't new concepts, but they are difficult for a brand to pull off. This is getting even more so as legal and compliance vultures take greater control over your social content and sharing. Humor creates these magical moments of connection and it's a big part of how we be human online.

Simply being nuts and bolts practical with you, you're going to be better off handing these keys off to User Generated Content coming from your Customer Community. The combination of effective brand advocacy and user generated content sharing has the potential to fuel rapid growth in a company with a zero paid media policy.

User generated content is all at once promotional, trustworthy, relevant, influential, and memorable.

Surveys indicate that 92% of social media users trust UGC more than traditional advertising when creators are not paid for those promotions.

Surveys also indicate that **84% of purchasing decisions of millennial buyers are specifically influence by UGC** and 35% of the same buyers found UGC content more memorable than any other form of media.

As you seek to boost sales by building engagement and connections through your consumers' own content, you will find that this same content will outperform your brand created content. You get the added benefit that consumers are eager to create and have their content featured by your brand.

Brands are often afraid to be considered 'pushy' so they don't provide clear instructions for creating and submitting content that both meets their guidelines and also allows customers to express themselves and influence others in the way they want to.

Marketing experts agree that visual media contests with photo or video submissions are effective strategies for gathering user generated content because they empower customers in their role as Brand Advocates.

Inside the Smync platform community, photo and video contests are already set up to be easily and automatically run to your community, and the winning submissions can easily be released for sharing both by the creator and the rest of the Advocate community.

When 85% of consumers say they find UGC more influential than brand created content, and it costs nothing, it might be time for you to listen. It is important for the future of your brand to have a proactive strategy for connecting with customer content creators, empowering them and leveraging their creations to increase the performance of your marketing department across all channels.

Here is a terrifying and weird example of the power of user generated content. 25% of all of the global social communication covering ISIS is created, directed and distributed by a special group of, oh well, I'll just call them 'super fans.' These Advocates are known by the highest levels of ISIS leadership and consistently recognized and rewarded by positive feedback for their contribution to the cause.

The Smync community platform allows users to load their own content creations to the platform, not just for special contests, but all the time. Then both of you have the opportunity to determine if that content is shareable from your brand owned community. Obviously, they are free to create and share their own content related to your brand right now. And they do it, completely free of your influence, without either your awareness or ability to leverage any of it for your benefit.

User generated content in the Smync community solves that problem immediately. Customer advocacy programs must be part of your overall customer experience strategy and brand marketing. Creating successful customer experiences, and embracing user generated content created by social Brand Advocates, gives you the opportunity to turn every customer into a lifelong Advocate.

Chapter 23
Your Movement

Think back to the Axis of Evil and you will see the disturbing trends. It is easy to see that your customers and prospects are overwhelmed and distracted. They're disconnected from your marketing attempts, intentionally. And, worst of all, they do not trust you!

Most prospects simply ignore you and your messages. They swipe through your Posts and advertisements. They skip your commercials or have turned off TV altogether (According to Google, a record 600,000 Americans disconnected from Television in 2015). They're all on the Do Not Call list. They've not been to their mailbox in a week. And the younger ones are abandoning established networks to escape your marketing and instead, engaging directing and exclusively with their friends on messaging services. In short, you've lost them.

But the omnipresent device in their hand and the six hours per day they spend connected to their friends and family on social media offers hope and opportunity! Because they DO trust each other and they ask each other's advice and then take action on the advice they receive.

They ask their friends where to eat, what to buy, what club to join, where to go and how to get there. They even seek the trusted advice of their social friends when it comes to the most significant questions of life; where and how to get a job, how to finance their dreams, whether to buy a home, get married, and everything they need to know about having and raising a child.
The beckoning opportunity for you is clearly via your empowered and fired up Social Brand Advocates. Those who are naturally, passionately, expertly, involved in your market and with your brand.

They know everything about you and your competition, and they have the greatest influence over whether the next generation of consumers buy your products and services. Will they be on your team or will you still be launching campaigns to target them for conversion?

There is overwhelming evidence that everything I say is true and that paid media will do nothing but continue to be more expensive and less effective. This includes the overly hot tactics of

Influencer and celebrity endorsement. You're wasting your money and your breath on tactics that your marketplace does not trust and will not only ignore but actively avoid.

Why pay Bieber $200K to post a Tweet when you could do something amazing with your actual customers and have 10,000 of them send an authentic Tweet about their love and affection for your brand?

Gratefully, as many marketers wake up to the wasted investments in mercenary Influencers, so also are many marketers waking up to the uselessness of vanity metrics. They are questioning the value of Likes and recognizing that while you can pay money to get Likes, you can't sell Likes for money, and you certainly can't put them on your Balance Sheet.

So, while your ideal audience is spending six hours a day on social media, and they have Liked your Page, less than 1% of them will ever be back to the Page or be engaged directly with your brand on any social network.

The other key fact that is finally being recognized is that social media is not, in fact, 'free marketing'. Organic reach is dead, it died in 2015, and it's not coming back.

To do social media as social organically and have it function naturally for a business brand, without the massive paid ad investment, would require an army of people and countless man-hours. And the messages would still be coming from a place that 96% of people think lacks integrity. But not doing social media is clearly not an option today, so what are you to do?

It is time to start mining the Gold you've been sitting on all along! Start mobilizing your authentic fans, empowered through effective social brand advocacy.

I am a stakeholder in the Smync social platform and I have explained and advocated for the amazing power of this tool. But I'm also an experienced business owner, and have been a Dow Industrials level corporate director.

I have never seen a more cost effective marketing approach than brand advocacy. As I look down the road and see the evolution of Live Social content and the explosion of dark social messaging apps like Snapchat, I see no other viable long-term marketing strategy.

You must remove the barriers between your company and your customers, allow them and empower them to use the device in their hand and liberate their innate desire to help others and their passionate interest in your subject and in your brand to fuel your marketing tomorrow and into the future.

I concur with Fred Reicheid, father of the Net Promoter Score analysis who said, "The only path to profitable growth may lie in the company's ability to get its loyal customers to become, in effect, its marketing department."

Your customers are already behaving in this way. I sat in awe the other day as I watched my wife grab her phone, open Facebook, and connect with her 767 Friends to ask for their advice on a home service and service provider to do a job. The response was nearly as fast as a Google search result, and came with personal insight and the advice of her trusted friends. The response even included a direct call to action link to one friend's highest recommendation. That link was pressed, that service was called and scheduled, and not a single type or source of any kind of traditional marketing or search tool was even considered. The whole transaction took perhaps five minutes. The deal was done.

The service provider did not have to provide one shred of evidence, produce an answer to a single objection, or spend a single nickel on marketing to make that sale happen. The customer experience of my wife's Friends, shared over Facebook at just the right time was all the information needed to complete the transaction. The future of marketing is here and that is what it looks like!

I know that your budgets are tight, you have existing relationships with vendors, planned campaigns, and all manner of people both passionately committed to and functionally assigned to executing your existing marketing strategy. Scrap it and start over. Take a hard look at that budget and look hard through those upcoming plans. Evaluate performance and value, and fire all underperforming tools immediately.

Carefully consider the realistic value of your planned, paid advertising campaigns as well, and seek things that could be drastically trimmed back. You'll have no trouble justifying these shifts. Invest instead for the long term in a social brand advocacy community developed by leveraging the Smync platform.

It is already likely that 95% or more of your transactions today are triggered by a word-of-mouth connection, like the one I just described, or an off-line conversation among friends. Why not invest in the most proven strategy powered by the wings of the greatest revolution in the history of human communication?

Make Social Brand Advocacy the significant focus of your Marketing Plan immediately.

How to Connect:

I am very grateful that you have invested your time to reach this point in the book!

I would be pleased to invest back into you and to connect directly to discuss how Social Brand Advocacy and Smync can help ignite your movement and grow your brand.

Below is a link to my personal calendar to schedule an appointment for a call:

http://bit.ly/ApptWithShawn

I would also be pleased to connect online.

You can find me on all Social Networks as *@ShawnMMiller*

theshawnmmiller

www.Smync.com

@GoSmync **on Social**

And at 1871 in Chicago

Special Credit and Thank You is due to **Jeff Ernst**, my LONGtime friend and the CEO of Smync, for igniting me on this mission and facilitating along the way. *Thank you, Jeff!*

Closing Note re: Purpose:

As I journeyed through the curation and production on this book and began also blogging, speaking, tweeting on behalf of Smync and Brand Advocacy there has been a clear ring in my heart.

That somehow the mindset and tactical shifts directed for successful Brand Advocacy would also fan a spark of reconciliation and ignite flames of greater movement of restoration of the massive distrust and division identified here as the problem in connecting with consumers.

Distrust and division in our society seems more profound than I have ever felt before.

- Consumers and Companies
- Employees and Employers
- Vendors and Customers

And so I leave you with one last drawing and my heartfelt encouragement to be the change you seek and to use your position as a leader in the societal pillar of commerce to reduce the division and build cooperation and coordination for the benefit of us all.

Special Credit and Thank You to **Kellan Fluckiger** for his wisdom, guidance, and encouragement through this whole book creation process. *Thank you, Kellan!*

www.ingramcontent.com/pod-product-compliance
Lightning Source LLC
Chambersburg PA
CBHW060031210326
41520CB00009B/1080